LET THERE BE LIGHT

THE STORY OF RURAL ELECTRIFICATION IN KENTUCKY

David Dick

First Edition, November 2008

Copyright by
Plum Lick Publishing, Incorporated
P. O. Box 68
North Middletown, KY 40357-0068
or
1101 Plum Lick Road
Paris, KY 40361

www.kyauthors.com

―――――――――――

Dust jacket design and book production by Stacey Freibert Design

Dust jacket artwork by Jackie Larkins

Photograph of David Dick by Chuck Perry

―――――――――――

Other books by David Dick

The View from Plum Lick
Peace at the Center
A Conversation with Peter P. Pence
The Quiet Kentuckians
The Scourges of Heaven
Follow the Storm: A Long Way Home
Jesse Stuart—The Heritage
A Journal for Lalie—Living Through Prostate Cancer

Books by David and Lalie Dick

Home Sweet Kentucky
Rivers of Kentucky
Kentucky—A State of Mind

ISBN: 0-9755037-3-1

EAN: 978-0-9755037-3-7

Library of Congress Control Number: 2008903199

To

J.K. Smith

*A leader who guided the
creation and maturity of the
Kentucky Association of Electric Cooperatives*

CONTENTS

PREFACE

Brothers and sisters, I want to tell you this.
The greatest thing on earth is
to have the love of God in your heart,
and the next greatest thing
is to have electricity in your house.

Farmer giving witness in a rural
Tennessee church in the early 1940s.
(*The Next Greatest Thing*, National Rural
Electric Cooperative Association)

Not so very long ago a great darkness covered much of the countryside of the United States. In about ninety percent of rural areas, there was no electricity.

"I was one of eleven kids raised on a fifty-acre farm in Ohio County, Kentucky, when the electricity came across Pumpkin Ridge within about seventy-five feet of our farm," says Naomi Ives of Brush Run Road in Jeffersontown, Kentucky. She speaks from her vantage point in the twenty-first century of light bulbs, washing machines, and running water inside the house where it's sorely needed and appreciated.

"We had Papa to tell us good ghost stories or play games with us around a big open fire. We had love and togetherness," recalls Naomi. But she, her parents, and her ten brothers and sisters were caught in a Catch-22. They knew they had a need but could not afford to fill it; the longer they waited, the deeper they plummeted. Their plight stretched across what was meant to be a commonwealth whose motto was "United We Stand, Divided We Fall."

Private, investor-owned power companies had certainly brought lights to towns, urban areas, and some nearby rural places, but the idea of reaching far out to isolated homes up the Pumpkin Ridges where mules were often the main transportation—well, that was not seen as part of the profit picture.

A new approach—rural electric cooperatives—became an urgent necessity. Cooperatives would go to work wherever profit making had failed. Near the end of the twentieth century, Touchstone Energy was incorporated as

In 1935 with one stroke of a pen, President Franklin D. Roosevelt paved the way for the establishment of the Rural Electrification Administration. (KAEC archives)

the national brand identity for electric cooperatives across the nation. This brand initiative has raised the public understanding of what makes co-ops different and valuable.

The late Kentucky historian laureate Thomas Dionysius Clark was acutely aware of the absence of electricity in the hills along the Kentucky River, widening out from its three forks and main stream.

"The erection of electric cooperatives across Kentucky was a monumental accomplishment," said Dr. Clark, "requiring extensive planning, rights-of-way acquisitions and the stringing of lines. The landscape would never again be devoid of these landmarks of modernity. These [power] lines trailed through hill country, up deep ravines, over rocky knobs and through cornfields and pastures to reach the doors of remote subscribers."

The undergirding included the enactment of the Tennessee Valley Authority Act—May 18, 1933—which, Dr. Clark concluded, "contained the seminal mandate of the broad social betterment of a region. Specifically, this meant the generation and distribution of cheap electric current without reference to urban-rural boundaries."

Dr. Clark had a sharp sense of rugged life in remote cabins where families coped for so long, waiting for the miracle of electricity. No church was immune to the darkness; no school or home could count on cooler days and warmer nights. It was as if there were two worlds—one bright with the simple throw of a switch; the other dark and harshly forbidding. There was a gross inequality crying out for correction.

The answer came with President Franklin D. Roosevelt's belief that there needed to be lights on the American farm.

On May 11, 1935, President Roosevelt signed Executive Order 7037 paving the way the following year for the establishment of the Rural Electrification Administration (REA). There was no turning back. Electric energy would dispel the darkness in rural America.

"This was followed," according to Dr. Clark, "with the enactment by Congress of the broad administrative law authorizing the creation of regional cooperatives. These cooperatives would generate and distribute electric current and service to the rural communities, no matter how isolated they might be.

"Signing the executive order on that hurried afternoon in 1935 was a cardinal step in the modification of rural life in the Republic during the era of the ravaging Great Depression. As eloquent as the text was, it hardly portended the spiritually moving moments when at last the lights were turned on for the first time in rural Kentucky and elsewhere.

"The organization of folk-managed cooperatives and the unification of common human energies must surely be one of the most dramatic demonstrations of the application of grass roots democracy in American history."

Farm families would trim their wicks and read Chapter One of Genesis—a stirring call for action:

And God said, Let there be light:
and there was light.
And God saw the light, that it was good:
And God divided the light from the darkness.

Let There Be Light—The Story of Rural Electrification in Kentucky recalls a need whose time for resolution was long overdue. It speaks with the voices of a resourceful people, recalls strong-minded leaders, and witnesses a rousing vision of cooperative effort.

One of these individuals was a twenty-one-year-old youth—J.K. Smith—who saw the possibilities of something as fundamental as fossil fuel, generation of electricity, poles set upon the hilly land, and lines extended to simple light bulbs where before there had been mostly candles and kerosene lanterns.

With courage and inspired leadership, J.K. made rural electric cooperatives a lifelong mission. He and those other rural electrification pioneers could hardly have imagined many of the consequences that the arrival of electricity would produce in rural Kentucky.

Husbands would no longer have to bring in crudely-cut timber for wood-burning stoves. Housewives would rejoice with the passing of "sad irons." Children could begin to dream about such delights as radio, television, and decades later, the wonderland of computers—even a bright light with which to read books.

David Dick
Plum Lick, Kentucky
November 1, 2008

FOREWORD

*We have to have some vision about where we're going
and how we're going to get there.
It all comes down to common sense:
observing, being up to date, staying on top of what's going on,
being aware, anticipating problems down the road
and trying to plan for them.
If you use common sense, you can do just about anything.*

J.K. Smith
Founding President
Kentucky Association of
Electric Cooperatives

A Conversation with J.K. Smith
March 16, 2007
Mt. Dora, Florida

"May I ask how old you are?"

"Ninety-one."

"Take us back to life on the farm in Meade County, Kentucky."

"It was in the early '30s. We had a two-hundred-acre farm, general farming. I was one of five children. I went to grade school in Ekron, Kentucky, high school in Brandenburg—Peewee Reese's town."

"What was it like in the daily work of your mother and father?"

"They were typical farmers. Had about fifteen cows. Had sheep. They'd kill about eight hogs a year and process the meat. Kill a beef each year. Canned vegetables."

"That must've been a challenge without electricity."

"We had a carbide plant for lights. You'd put this stuff in a big hopper, put water with it, and it'd create gas and you'd get lights."

"How strong was the light?"

"Fair. Just fair. Had to use it sparingly. Nobody complained. We had to have it."

"Was it typical to have that kind of light?"

"No, it was not."

"How did your father know about it?"

"Don't know, but he put it into the basement in great big tanks, and when you got the residue out you made whitewash and whitewashed the fences with it. It was lime stuff.

"I'd be out on a date and come home to milk the cows on a Sunday night. Used a lantern. We'd separate the milk with an old hand-operated separator. We'd have an icehouse square hole in the ground with a cover over it. Go out in the morning to get some to put in the wooden refrigerator box. Used a pond with no stock around it, pretty well protected. Had a cistern with a hand pump—one cistern in the side yard, one cistern in the back yard. Had a wood stove in the kitchen with three or four irons. Hung clothes outside on the line to dry."

"Farm equipment?"

"Old Ford tractor. Very few farms had one. If you had a load on it, it'd rare up on you. Very dangerous."

"How'd you get the message that it was time for a better way of doing things?"

"Well, I went to school in Bowling Green and at that time it was called Western Kentucky Teacher's College. That was in 1936. Went two years there and in the summertime I'd come back home and in one of the summers, I had a brother who was an attorney, Henry Smith, and he was attorney for the development of a Meade County rural electric system.

"So the two of us got to working in the summertime for the contractor out of Louisville. The engineer was Ray Chanaberry. He was the engineer for the rural electric systems."

"You didn't know anything about the mechanics of it?"

"No, I knew nothing about it. I picked it up from there on."

"Did you have a vision?"

"I could see it developing all over the state. I was one of the early ones. Chanaberry was the engineer for most of the systems at that time. He told us about one being developed in Flemingsburg. He was with the engineering firm on the job. They were looking for a superintendent. I suggested that I was interested. I was twenty-one years old. I picked up the nomenclature, a little bit of the procedures from the one in Brandenburg."

"Mr. Smith, as I think about it, it all seems complicated—so difficult to supervise the building, the hiring, the getting of good people, the satisfying all the requirements of the Rural Electrification Administration."

"I could see it developing all over the state. I was one of the early ones," says J.K. Smith, the visionary whose time had come. (KAEC archives)

"I don't know how I did it either. They were foolish to try it with me. I went up there to the Board. My brother went with me, and I was interviewed by the Board as an applicant for superintendent, which they had to have before they could get their money advanced from REA— $200,000 to build 600 miles of electric lines in Fleming and Mason counties. I was one of nine applicants, and I think by virtue of the fact that I knew some of the nomenclature, knew a little bit about the procedure, the Board was impressed enough to give me a shot at it."

"How much did they pay you to start?"

"Seventy-five dollars a month. I was making eighteen when I got married."

"What year were you married?"

"February 12, 1938. Been married sixty-nine years. Met Marge in Bowling Green. I like to tell people that she's the best thing I got out of college."

"How long did you stay in Fleming?"

"I started to work there August of 1938, and I stayed there until 1948. I went on to Kentucky Association of Rural Electric, and I started it. Had no staff, just a lot of big ideas."

"Why did you think it necessary to have such an association?"

"I could see from the standpoint of my ten years at Flemingsburg that there were so many things that they needed—to get out information about the program, for one thing. No one understood the program. The members needed the information, as well as the general public.

"We were going to be facing problems with the legislature. KARE— when I started it was called Kentucky Rural Electric. I was the first General Manager. They had no staff, very little funding. Each system had to put up the membership money."

"When you were there [Kentucky Association of Electric Cooperatives] did you have to be careful about not stepping on the toes of the cooperatives?"

"KAEC is the organ to hold them together, the twine to hold them together. Money was scarce. We told them we'd pay them back with materials—utility wires, transformers. Each electric system is responsible, each is autonomous. We started out doing the billing. Members were reading their own meters. We'd transfer that to a meter book. Ninety-eight percent of them were honest.

"CFC [National Rural Utilities Cooperative Finance Corporation]. That was started in 1970. It's a banking institution. I foresaw that money was going to stop coming from Washington. Those of us in Kentucky were pushing that we needed to do something for the national shortage of money. I had the idea but I didn't know how to put it together. Wall Street people (investment banker Kuhn Loeb Company in New York) had a very sympathetic view of rural electricity. First we had to convince the federal government to share the mortgage with us. That was a terrible job."

J.K. Smith remembers, "I was one of the originators of East Kentucky Power. Each co-op was buying on a wholesale basis from their own areas. That was sixty percent of the costs. We were buying from Kentucky Utilities. We went to the Public Service Commission, had hearings. It was very political.

"I happened to be in the program when I could do a lot of pioneering. I could see the needs, what was coming ahead, and I had the gumption to try it. I guess exposure to different people keeps your mind open to new ideas. If you have a need, go after it. You have to be a good salesman.

"One of the interesting things we did in Kentucky was to have a national rural electric program with other countries—Ecuador and the Dominican Republic. We supplied materials that we'd outgrown.

"Efficiency is the main thing—mergers—you've got to have enough income to cover expenses. You need certain leaders who you can depend on, seven or eight managers who were your key people. You start talking to them first.

"This is not a farm situation anymore. If we got three users per mile, we thought we were doing well. Now, cities are expanding and we had fights with the cities over territorial protection. We had very little commercial use. You've got to anticipate the future. You've got to be ahead of it."

"If you were writing this book, where would you begin?"

"Start with people—it's a people program."

THE BEGINNING

You are on the way to a better life. My Old Kentucky Home
will have a new meaning in the future to thousands.

John Carmody
Head of the Rural Electrification
Administration
Washington, D.C.
May 19, 1937

"Normally early risers in Henderson County, Kentucky, where the
Green River empties into the Ohio, awoke even earlier that morning,
May 19, 1937, in order to finish their chores and get into town. These
hard working people rarely left their farms in the middle of the week, but
this was a special day."

This previously unpublished "The Beginning" was written and
contributed by Gary Luhr, former editor of *The Rural Kentuckian* and its
successor, *Kentucky Living*, publications of KAEC—Kentucky
Association of Electric Cooperatives.

"It was Pole-Raising Day when the government's two-year-old
program to bring electricity to rural areas took visible form in Kentucky.

"Almost a year had passed since men of vision, who knew that
electricity could relieve the drudgeries of rural life, had formed the
Henderson County Rural Electrification Association. They had spent the
year traveling over what passed for country roads in those days, signing
up friends and neighbors who were willing to pay five dollars to have
electricity installed in their homes and on their farms. The Henderson
County Fiscal Court had granted the association a franchise in July 1936,
and by October, the Rural Electrification Administration (REA) in
Washington had approved a $190,000 loan to build the first one hundred
and fifty-three miles of line. By fall 1937, approximately one thousand

1

two hundred people would have electricity for the first time.

"The air that morning had its usual spring chill. Fog hung in patches along the creek banks as farmers and their families made their way by wagon and car toward Henderson. In town, the stores opened for business as usual, but merchants planned to close them at noon in order to join in the day's special event. Schools in the county would also be closed.

"On the north side of town, volunteers had decorated a stage and erected loudspeakers in the middle of Atkinson Park. Radio station WGBF from across the river in Evansville had installed equipment to broadcast the proceedings live that afternoon. By mid-morning, the air was heavy with the sweet smell of barbecuing mutton, which the Henderson County Homemakers planned to start serving at noon. At 10:30 a.m., Director Otto Geiss signaled a downbeat and the Henderson County High School Band began a thirty-minute concert. Mayor L.L. Hurley and County Judge R.R. Crafton welcomed the crowd, which by noon totaled around three thousand.

"Half a century had passed since Thomas Edison ushered in the electric age by perfecting the incandescent lamp in 1878 and by building the country's first electrical generating plant in Lower Manhattan in 1882. By the 1930s, most cities had electricity. The city of Henderson, Kentucky, had seen its first electric lights in the summer of 1886. Electric utilities were a $12 billion industry in the United States, but one that existed almost exclusively within the boundaries of cities and towns. Utility executives perceived the cost of extending electric lines into sparsely settled areas as prohibitive. As a result, by 1935, only ten percent of Kentuckians received electricity from a central source.

"The few farms that had electricity did so because they lay along main roads, where the power companies ran their lines, or because the farmers had paid to have the lines extended. Some farmers had battery systems run by gasoline, which produced enough electricity for lighting but little else. Most farmers in Kentucky had neither, however. They were the ones who had gathered in Atkinson Park in the spring of 1937.

"The main speaker that day, Governor A.B. 'Happy' Chandler, had grown up on a farm about a dozen miles away. He had come to

Henderson at the invitation of his boyhood friend, Ben Niles, President
of the Kentucky Farm Bureau. Niles had been instrumental in forming
the Henderson County Rural Electrification Association. One by one that
afternoon he ushered speakers to the microphone: E.J. Kilpatrick of the
University of Kentucky Cooperative Extension Service; T.B. McGregor,
a member of the Kentucky Public Service Commission; and Ben
Kilgore, executive secretary of the Kentucky Farm Bureau and a strong
crusader for rural electrification.

"Kilgore told the crowd electricity would be one of the greatest
contributions ever to the farm people of Kentucky. John Carmody, head
of the REA in Washington, sent a telegram saying, 'You are on the way
to a better life. My Old Kentucky Home will have a new meaning in the
future to thousands.'

*"Pole-raising Day" in Atkinson Park, May 19, 1937, Henderson County, Kentucky.
Taps was blown, a flag was raised, and a funeral held to bury the coal oil lamp
and raise a new cooperative electric power pole of progress.* (KAEC archives)

"Just before noon, twelve teenage girls from the Smile Awhile 4-H Club at nearby Hebbardsville proceeded solemnly to where the first pole was about to be raised. They wore white dresses, and some carried white dogwood wreaths. For days they had practiced how to move and where to stand. Attention focused quickly on the four who carried a small wooden box. Inside lay a glass lamp about to be buried as a symbol of a lifestyle about to pass.

"Henrietta Priest, whose father had built the coffin, led the girls in reciting the 4-H pledge. As the lamp was lowered into the ground, she pronounced the day's theme: 'From Darkness into Light.'

"Around 2 p.m. the association's attorney, F.J. Pentecost, introduced Chandler. The Governor said he was glad Kentucky's pioneer effort in rural electrification had occurred in his home county. Electricity, he said, would take farm women out of the 'drudgery class.' As he concluded his remarks, a small group of men set the first wooden pole into the ground. The pole was wrapped with red, white and blue lights, and when it was in place, Pentecost flipped a switch that turned them on.

"The crowd applauded. Many cheered. A few cried.

"The band played the *Star-Spangled Banner* while a group of Boy Scouts raised an American flag to the top of the pole....For those who were there the pole itself had become a new symbol of freedom."

From West Kentucky and Hickman-Fulton, to Big Sandy and Grayson RECCs, from Owen and Fleming-Mason to South Kentucky and Cumberland Valley RECCs, and eighteen other electric cooperatives, the story of rural electrification in Kentucky reads like a dream—a vision of lights—inspiring men, women, and children to see a better, brighter day.

As J.K. Smith so wisely stated, looking back from his ninety-one-year perspective: "I happened to be in the program when I could do a lot of pioneering. I could see the needs, what was coming ahead, and I had the gumption to try it. I guess exposure to different people keeps your mind open to new ideas.

"If you have a need, go after it."

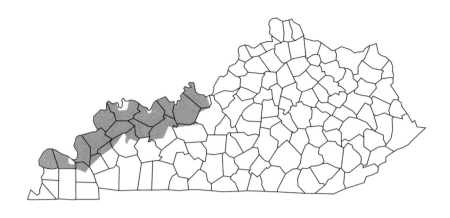

BIG RIVERS
ELECTRIC CORPORATION

*We're all together in the Kentucky Association of
Electric Cooperatives (KAEC), so it shows that we can all
work together, which is the key to our success.*

Mike Core
President & CEO
Big Rivers
Electric Corporation

Big Rivers was created in 1961 by community leaders from the
Green River, Henderson-Union, and Meade County cooperatives. They
positioned themselves within a strong circuit of new hope, new
efficiencies, and a fervent belief that they could generate their own
wholesale electricity. The founders of Big Rivers would assume
responsibility for generating, then transmitting power to keep the lights
on and promote economic development in rural areas, until then cold-
shouldered or factored out of profit-building strategies by entrenched
investors.

Private power companies vigorously fought what they saw as the upstart Big Rivers plan, dismissing it as "The Great Give-Away," citing the two percent loans from the federal Rural Electrification Administration as unfair and detrimental to the concept of free enterprise, an argument that would persist for decades.

The idea of investor-owned utilities and cooperative governmental initiatives working side by side was slow to take credible, lasting shape. But rural Kentuckians in the 1930s had grown tired of gnarled and blistered hands pulling up buckets of deep well water in all seasons of the year, groping to the outhouse in February, bone-tired of milking cows with cold and aching fingers by the flickering light of kerosene lamps. Farmers considered themselves lucky if oil lanterns didn't turn over and barns didn't burn down. Long-suffering housewives and their daughters had wondered if there might—just might—be something better than "sad" irons to press their husbands' and brothers' Sunday pants. The sleight of hand was to moisten the fingers and spank the bottom of the iron to check to see how hot it was before proceeding with a "sad" test of stern-eyed willpower.

It was time to throw a new switch in the competitive world of electrification.

J.R. Miller and Robert A. Reid Sr. of Green River Electric, John Hardin and Robert Green of Henderson-Union RECC, and Kenneth Coleman and Leslie Jenkins of Meade County RECC met to formulate plans to construct a generation and transmission cooperative that would serve the wholesale needs of their respective cooperatives. By 2007, with the addition of Jackson Purchase Energy Corporation in 1977, the number of customers reached 110,000 along a thousand miles of transmission lines. The planned objective of this part of grassroots Kentucky was simple: maintain the lights along the headwater tributaries from Wolf Creek in Meade County to the Tradewater River in Union and Crittenden counties, and eventually westward to the Mississippi River, where Jackson Purchase Energy Corporation was organized in 1937. This co-op initially received its wholesale energy from an investor-owned utility, Kentucky Utilities, but later joined Kenergy (a consolidation of Henderson-Union and Green River co-ops) and Meade County in the operation of Big Rivers Electric.

J.R. Miller played a prime, passionate role in the ambitious, some might decry audacious, founding of Big Rivers. As Democratic Party State Chairman and General Manager of Green River RECC, Miller— intense dark eyes brooding beneath his leonine forehead—was said to be admired and feared, scorned and courted, a man of relentless energy, a man who understood and sought power—an absolute ramrod, a genius for organizing. Miller was Chairman of the first Big Rivers meeting and served as a Board member until 1975. He retired from Green River Electric after thirty-five years of unceasing service.

J.R. Miller and his pride of associates began the time-consuming, uphill challenge of obtaining a Rural Electrification Administration [REA] loan and Kentucky Public Service Commission approval to construct the new power plant.

J.R. Miller's son, Jim, Corporate Counsel for Big Rivers, in 2008 recalled how his father made his way through the bureaucratic levels in Washington, D.C.: "During the course of the efforts to form Big Rivers Rural Electric Cooperative Corporation (subsequently renamed Big Rivers Electric Corporation), the loan to fund that effort and construction of the first Big Rivers generating station became bogged down at REA. When he was unable to move REA, J.R. turned to his friend and former U.S. Senator, Earle Clements, for advice. Clements arranged for a meeting with President John F. Kennedy on the subject. Clements and J.R. met the President in the Oval Office, and after exchanging pleasantries (which should not have taken long since J.R. was a delegate committed to Lyndon Johnson at the Democratic National Convention in 1960, where Kennedy became the eventual nominee) they explained their

J.R. Miller played a prime, passionate role in the ambitious, some might decry audacious, founding of Big Rivers.
(Big Rivers EC archives)

7

problem to him. President Kennedy picked up his telephone, called Orville Freeman, his Secretary of Agriculture, and asked if there was any reason why the Big Rivers loan could not be approved that afternoon. Apparently there was none, as the loan was approved by the end of the day."

The Board received REA approval on July 6, 1962, for an $18 million loan to construct a 65,000 kW plant. Groundbreaking ceremonies were held October 23, 1963.

The first Big Rivers generating installation—the Robert A. Reid plant—was completed in 1965 and became operational on January 1, 1966. Those were the heady days of increased industrial power demands—Anaconda Aluminum, National Southwire, and the National Aluminum Company were attracted to the newly minted possibilities of western Kentucky. There was a newly arrived dynamic of economic growth—jobs, payrolls, and a spurt of sales in real estate fueled by redesigned agricultural machinery and a parade of other coveted consumer products—most important, though, vastly improved life on the farm.

And so was born Big Rivers.

Rural Electrification in August 1967 noted that "Nobody in Kentucky is speaking louder with better results than J.R. Miller and Green River RECC. An aggressive and imaginative development program has attracted $155 million in industrial investment to Western Kentucky since Big Rivers Generating Plant began supplying the co-op low-cost power in January 1966. J.R. Miller, the dynamic manager of Green River, firmly believed rural electric co-ops should accept community responsibility and assume leadership in promoting the welfare of citizens within their service area."

From its birth in the early '60s to the present, Big Rivers has survived a rate increase controversy, internal scandal, and the jungle of bankruptcy. The decade of the 1980s reads like a whirligig spinning through the Ohio Valley, with bits and pieces of fractured forecasts left along what had normally been the optimistic way.

Ground was broken for the second unit at the D.B. Wilson plant in

Ohio County in the Western Kentucky Coal Field but had to be dropped when the national economy turned downward in 1982-1983. Industrial customers, especially the aluminum industry, experienced a severe market decline and the federal government canceled two planned synthetic fuel plants.

Big Rivers requested rate increases to help pay for the D.B. Wilson plant, but the Public Service Commission delayed a decision. Aluminum companies protested and talked about closures. Big Rivers responded with its own fears of bankruptcy; it defaulted on a $20 million debt service to the REA. The U.S. Justice Department sued for foreclosure then approved a settlement.

Big Rivers offered the Kentucky Public Service Commission a revised rate/debt plan. Big Rivers owed $1.3 billion in loans provided by REA, including $756 million to build the Wilson plant. The dream of J.R. Miller and other rural electric leaders of western Kentucky would be debated again and again.

"Big Rivers now at full capacity" was the headline in January 1987.

Scandal is every organization's worst foreboding. James Miller, Big Rivers Corporate

The Robert A. Reid /Robert D. Green plants began with one installation, the Reid plant, for which ground was broken October 23, 1963. On-line January 1, 1966, it generated 65 megawatts, but now, combined with the Robert D. Green plant, they generate more than 580 megawatts of power.
(Big Rivers EC archives)

9

The D.B. Wilson plant in Ohio County was named for the long-time Meade County RECC Board member of the same name. Shown here under construction, ground was broken for the plant in June 1980 and it was completed in 1983. The plant now generates 420 megawatts. (Big Rivers EC archives)

Counsel, son of J.R., explains the case of a former top executive at Big Rivers:

"William H. Thorpe, former General Manager of Big Rivers, was indicted in 1994 and convicted in 1996 for violation of several federal laws in connection with a scheme by which he supplied certain coal vendors information from confidential bids by other coal vendors for Big Rivers' coal supplies. On December 16, 1996, he received a sentence of fifty-seven months in federal custody, plus a $50,000 fine, plus restitution to Big Rivers of $3,189,537.81."

Bankruptcy is a legal resolution of financial troubles. Again, James Miller:

"On September 25, 1996, Big Rivers filed a voluntary petition for relief under Chapter 11 of the United States Bankruptcy Code. Big

Rivers filed a plan of reorganization based on leasing its generating facilities to a subsidiary of PacifiCorp Holdings, Inc. for a period of approximately twenty-five years. During that period PacifiCorp would sell back to Big Rivers a substantial portion of Big Rivers' wholesale power requirements. In February of 1997, the bankruptcy court ordered an auction process to assure that the proposed PacifiCorp transaction produced the maximum value for creditors. In March of 1997, the bankruptcy court declared Louisville Gas and Electric the successful bidder in the auction, and in June of 1997, the bankruptcy court approved a modified plan of reorganization that essentially substituted LG&E Energy Corp. for PacifiCorp." The plan of reorganization, which included the transaction with LG&E Energy Corp., was consummated as of July 15, 1998.

"Big Rivers is working to take back direct management and operation of the generation plants that were leased to subsidiaries of what was then called LG&E/Western Energy Corp. (now E.On U.S.) in 1998. Should the 'unwind' of the lease agreements be finalized, Big Rivers will again operate and distribute the power produced by these plants, instead of buying it from an E.ON U.S. entity.

"It is vital that Big Rivers and its members have control of their own operations so that better decisions can be made about how best to serve the people of western Kentucky.

"From the mid-60s to the mid-70s, the growth within the Big Rivers system has averaged more than 10 percent a year for residential, farm, and small commercial consumers, not including much greater increases in large industrial loads like the Willamette paper mill, National Southwire Aluminum, Commonwealth Aluminum, Anaconda, Southwire Rod and Cable, and others.

"Approvals were obtained, ground was broken for the D.B. Wilson plant on June 20, 1980, and construction began.

"During 1982 and '83, the national economy turned sharply downward and the aluminum industry was devastated by a drastic drop in demand and prices. Both smelters closed one-third of their capacity and suffered large financial losses.

"During this time we survived because of intersystem sales, and the smelters continued to pay contracted demand charges.

"By late 1983, the economy improved, the smelters returned to full production, and our load was back to normal.

"Early that year we started preparing a rate increase application to pay for the Wilson Plant. Recognizing that an increase based on normal accounting methods would

Among the many who were on-site of the D.B. Wilson plant during its three-year construction were these four men, three of whom were Morton Henshaw, Harvey Sanders, and Glyn Truitt. (Big Rivers EC archives)

be a financial shock to consumers, we began negotiations for a sale-leaseback arrangement with General Electric Credit Corporation. This leveraged-lease was expected to save consumers over $700 million during the life of the plant. However, the lease was unable to be transacted."

Mike Core, President, CEO since 1997, and cooperative realist: "Most generating and transmission co-ops were formed to provide power supply, basic needs. Big Rivers was more specialized, put together to provide an economic development, unique in that it was formed to attract industry—we've been wildly successful.

"After WWII up to mid-70s, we'd had a bit of recession—we got caught by an Arab oil embargo which caused us to forever think about energy and its sources.

"As to the future, it will be more of a challenge to use coal, but the real issue is CO_2 and global warming—greenhouse gases. I'm concerned that we cannot change our energy policy overnight. It'll take twenty-five to thirty years to turn this ship. In the meantime, we've got to use the energy more efficiently.

"Bill Thorpe didn't put us in bankruptcy—that came as a result of building the Wilson station," says Mike Core. "The high costs of construction put a level of debt on Big Rivers that we couldn't sustain. But we have worked hard to better our financial condition, and we continue to improve."

At present, J.R. Miller's "not so wild a dream" Big Rivers has become an economic giant of western Kentucky—the Kenneth C. Coleman Plant (three units), Robert D. Green Plant (two units), Robert A. Reid Plant (one unit), and the D.B. Wilson Plant (one unit). Also, Big Rivers members have additional resources

The Kenneth Coleman plant, begun in the spring of 1967 in Hancock County, was completed on August 1, 1969, one month ahead of schedule. Here, a generator for the plant nears completion for the plant that now generates 455 megawatts of power. (Big Rivers EC archives)

available from the Southeastern Power Administration and a contract with the City of Henderson for its Station Two—located at the Big Rivers Reid-Green generating station.

In 1999, Green River and Henderson-Union were consolidated within Kenergy, while Meade RECC remained a cooperative standing on its own. These two organizations, along with Jackson Purchase Energy Corporation, receive their power from Big Rivers.

Total power capacity available to Big Rivers in 2008 is 1,854 megawatts or 1,854,000,000 watts—almost two billion.

"The future is in the hands of management and the cooperative owners, the consumers of Big Rivers," says Mike Core. "Ideally, each customer-owner will become actively involved. They will make their voices be heard to be sure that the work of the cooperative remains true to its mission."

From Big Rivers in the west to Big Sandy in the east, the cooperative movement has bridged the troubled waters of rural electrification.

BIG RIVERS ELECTRIC CORPORATION
ADMINISTRATION

William Rumans	1964 – 1975
Scott Reed	1975 – 1976
Marshall Dorsey	1976 – 1978
W.H. Thorpe	1978 – 1992
Paul Schmitz	1992 – 1995
Al Robison*/Scott Reed	1995 – 1997
Mike Core	1997 – present

(*) Al Robison not an employee of Big Rivers - worked by consulting agreement.

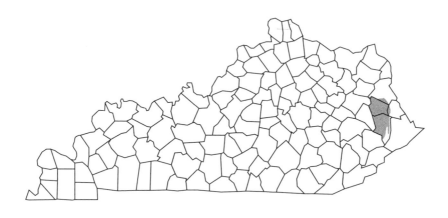

BIG SANDY
RURAL ELECTRIC COOPERATIVE
CORPORATION

*We started in '40, but we had a lot of people who didn't
get electric until the late '50s and early '60s.*

Bobby Sexton
President and General Manager
Big Sandy Rural Electric
Cooperative Corporation

The mountains and valleys of the Tug and Levisa forks of Big Sandy
will not return to an unlighted way. What profit-driven electrical power
companies have shied away from, the co-ops will go extra miles to reach
the farthest need.

The Big Sandy River from Catlettsburg to Louisa is a distance of
only 27 oftentimes muddy miles. The mountains and valleys along the
Tug Fork of Big Sandy trace for 154 miles the easternmost boundary of
Kentucky, while the Levisa Fork loops, twists, and sometimes thunders
164 miles back to Russell Fork and Fishtrap Lake in southern Pike
County. Tug and Levisa meet at Louisa. The land they embrace looms
larger than the combined states of Delaware and Rhode Island.

In her bicentennial book *The Big Sandy*, Western Kentucky
University historian and educator Carol Crowe Carrico recalls the legend
of how the eastern boundary of Kentucky was established along the Tug
and not the Levisa. Surveyors from Virginia were sent out with
instructions to visit both forks, and whichever was the stronger stream
would be the new state line. They came to the Tug Fork first, where a
storm rolled through, convincing them that it must be their choice. There
was a celebration that night, and the surveyors got quite drunk. The
storm moved west to the Levisa and next morning the surveyors realized
this stream was, in fact, probably the greater in volume of water. Yet
they felt obliged to confirm while sober that which they had decided
while drunk. Had it been otherwise, the future Kentucky might have lost
all or parts of Lawrence, Martin, Johnson, and Pike counties.

No matter the validity of the tale, the reality is a land mass of what
once was more than 4,000 square miles of candles, kerosene lamps,
and dreams of better lighted nights. Electrification through the remotest,
hogback mountains and deepest, fragile, and vulnerable valleys became
a job for rugged individuals with determination and stonewall grit. It
would require the vision of something as basic as cooperative effort.
Profit and dividends would be measured differently than the practice of
private power companies; nonprofit teamwork and helping hands would
be instrumental in lighting the cooperative way.

Big Sandy RECC was organized in 1940, eventually serving all or
parts of Breathitt, Floyd, Johnson, Knott, Lawrence, Magoffin, Martin,
and Morgan counties. As of 2007, there were 5,864 co-op customers in
Floyd and 5,461 in Johnson. Paintsville, county seat of Johnson, became
the location of Big Sandy RECC's headquarters—504 Eleventh Street. A
branch office would be based upstream on the Levisa, in Prestonsburg.

Reaching out from these locations are more than 1,000 miles
of electric line, about the distance from Louisville to Miami. But there's
more to Big Sandy Cooperative than poles, spools of wire, cranes,
trucks, and meters. There's a community connection under the
leadership provided by Bobby Sexton and a seven-person Board of
Directors.

"We started in '40, but we had a lot of people who didn't get electric until the late '50s and early '60s," says Bobby.

"Coal as *the* source of energy?"

"They say we've got one hundred years left, but that's not true unless we have different technology. It's not going to be as cheap as it has been in the past, but it's going to be cheaper than gas and oil. Every one of the counties we serve produces coal, Martin, Floyd, and Knott are probably the biggest producers of coal—more than one mine sends out a trainload every day. Pike County is the number one producer in the state. A lot of coal comes out of Jenkins." American Electric Power and Kentucky Utilities, two investor-owned utilities, serve parts of southeastern Kentucky.

Mountaintop removal is an ethical riddle, unthinkable for some, perceived as necessary and inevitable by others. In either case, it will not be ignored. Pressures are mounting from both sides, from within and without. The situation is potentially as volatile as it became during the strip mine controversy of the 1950s.

"Everyone has their opinion about that [mountaintop removal]. The only ones upset about it are

Early linemen were probably farmers and most assuredly unskilled in getting lines up on poles. A primitive affair at best. As electric conveyance has become more complex, linemen have become highly trained, skilled professionals. (KAEC archives)

17

the outsiders," says Bobby Sexton. In fact, there have been many passionate complaints from local residents who point to fouled water supplies and ravished vistas requiring years of reclamation. Yet, it is also true that one man's sanctified romantic mountaintop is another man's unromantic, less-than-holy source of opportunity.

"We welcome the level land. That's a rare commodity here," says Sexton. "We could go see one of the prettiest golf courses in the country in Floyd County built on some of that land. Our airport is built in Martin County. Every one of the counties we serve produces coal.

"Strip-mining got its bad name in the late '60s and early '70s during the coal boom, and we had some people who would get it any way they could. One of the wealthiest men in this county probably made more money than anybody in the state of Kentucky, but he was one of the main causes for the reclamation laws that went into effect."

"The future of Big Sandy RECC?"

"We're real upbeat. We're upgrading our co-op. We're doing

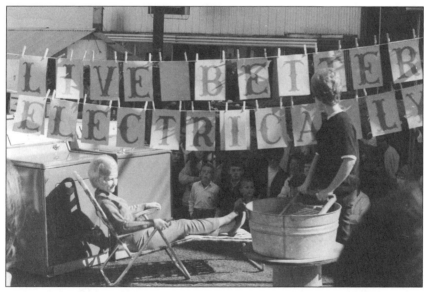

An Apple Festival parade is worth a thousand words—Goodbye washboard, hello "Give Mom a break!" seems to be the theme of this Big Sandy RECC float.
(Big Sandy RECC archives)

automatic meter-reading now. We still keep our prices low, and as far as Big Sandy is concerned, our prices may be a bit higher, but our service is ten times better [than private companies]. The current cost of living is going out of sight—our cost for the fleet has doubled. We have no plans to increase, but East Kentucky Power sets the [wholesale] rates. We're one of the sixteen co-ops for EKP, which has one of the cleanest, environmentally safe, coal-fired power plants in the U.S.

"We started a Kentucky Heritage Program. We wanted a lot of our students to know how we evolved and we invite members who are talented in soap making, corn shuck dolls, etc., to visit the schools and 'show and tell' the students. We also try to push the heritage of music, how it went from the mountains—pushing our history and heritage as well as we can. We give away six scholarships, a $1,000 scholarship in each high school."

Big Sandy RECC holds its annual meeting in the Mountain Arts Center in Prestonsburg, a facility strongly supported by the rural electric cooperative. Appearances at the Mountain Arts Center have resonated with Montgomery Gentry, Merle Haggard, George Jones, Loretta Lynn, Dwight Yoakum, and gospel groups, rock and roll groups, and Big Bands. The Mountain Arts Center's diverse educational programming has been attended by more than 200,000 school children from throughout a five-state region.

These are some of the beneficial consequences of one Appalachian electric co-op built upon a heritage and underpinning of self-sustaining pioneer stock.

"Then there is Bruce Aaron Davis," says Bobby Sexton.

"In the 'Read Around' program, Bruce Davis goes and reads to the kids in their schools. He's also a fine musician—mandolin. He does a lot of the music part of the heritage program."

Bruce is also known to play banjo, a sound that starts feet to tap, faces to smile, hands to come together. Music of Big Sandy is rich with variety, vibrating from Loretta Lynn's *Coal Miner's Daughter* in Johnson County's Butcher Holler, to Perry County's Jean Ritchie and her *Sorrow in the Wind*.

A visit to the Big Sandy office brings back memories undimmed in the daily lives of co-op members stopping by to pay bills, asking questions, saying hello to old friends, giving thanks.

Neva Davis Rowland hasn't missed a Big Sandy annual meeting since she started going. "When I was twelve, I rode to my first meeting in the back end of an International pickup," she says. So far she's attended fifty-nine meetings. She was seventy-two on March 17, 2008.

"I'm still using a 1949 Crosby refrigerator that belonged to my mother- and father-in-law."

Neva quietly, shyly remembers, "We burnt the coal oil lamps. Don't remember when we got electricity because the people above us wouldn't let the electric come over his land. We lived below and that kept us from getting it. We had a grocery store—that's how we got our first refrigerator. Father was mostly a tobacco farmer. We had no electric in the stripping room, we worked in an open barn. Each of us had a [stripping] grade—five grades. Coal stove in the kitchen and it was hot in the summertime—no fans, just screen doors—had screens over the windows.

"We heated the irons on the coal stove—kept three—we'd have two on the stove and iron with the other one."

"Water?"

"Drew it out of a draw well. That was a big problem too, because we had to go over a hundred feet to draw it. We had a pulley with a chain on the well. The water had a lot of iron in it. We had to wash clothes in Hood Creek—Lawrence County. I live in Lawrence County [in the area of Raccoon, Bear, and Tarklin branches]—born and raised there. When I married, I didn't go very far down the road.

"When you try to study by coal-oil lamp—I was just small then—it wasn't very easy."

Author and educator Linda Scott DeRosier grew up on Greasy Creek in eastern Johnson County, and she wrote about it in her book *Creeker*, living proof that Appalachian roots are as deep as they are promising, never mind the stereotypes perpetuated by Al Capp and his L'il Abner, Mammy, Pappy, Daisy Mae, and Moonbeam McSwine.

Chet Auxier, born up around Redbush in Johnson County (Wolfpen Branch, Patoker Branch, and Upper Laurel Creek), is getting on considerably in years, but he's got time to talk awhile with a writer passing through. Chet doesn't hurry his words, no need to sound like those lowland jabberwockers.

Beauty pageants tell a story of grace, poise, and a smile where once there might have been a non-electrical frown when having to iron an evening dress with a "sad" iron. (Big Sandy RECC archives)

Back in '39 or '40, "I spotted poles. Everybody read their own meters. My mother was afraid of it [electricity]. We had coal oil lamps...also had an Aladdin lamp. When electricity came through, oil was twenty-five cents a barrel—kerosene, my brother made his own kerosene—come from oil—had oil wells out at Fish Trap where they lived."

"Anything else remembered about pole setting in the time before the lights went on?"

"Cold mornings! If we got there a little late, the ground would start thawing—the oxen's feet would start sliding. We were paid one dollar a pole—twenty-five poles a day, but we got around fifteen in the wintertime.

"Built fires, heated water in wash tubs. Heated the water outside. All used the same water...oldest bathed first. Had twelve brothers and sisters—five boys and seven girls—we'd take turns about—sometimes

Substations provide a stepping-down point in the process of receiving power, reducing electric current strength, and sending it on to consumers. This is one of eight such substations in the Big Sandy RECC service area. (Big Sandy RECC archives)

we'd use the same water, sometimes we'd have to change water. Once a week. Now it's every day. There were big families back then to help raise the farm.

"We made our lye soap. Burn you up. Listened to radio on a battery. Used candles some. Stove wood to cook with...mixed it with coal to make it burn. Didn't go under the hill and mine much...they didn't need us. Pulled coal out of the mine with a wash tub...mined it for house coal.

"Hardest thing I ever did was logging—took a crosscut saw, had mules to haul the logs down to the mill to saw them into lumber. I was a driller in a saw mill. Run on gasoline...didn't start drilling until late '40s. Before that, I was growing up on the farm. Twenty head of milk cattle and had to milk them all by hand. Twice a day. Two hours each time. I got to where I could milk them first rate—that was in the '30s and '40s. My brother, Carson, he had the dairy—we hauled the milk into Paintsville.

"Went to work in the oil fields after the poles. We had to move from Fish Trap where Paintsville Lake is now. We logged, floated logs down Paint Creek...Sank a double-bitted ax in my foot. Horse and mule— didn't know what a tractor was...hardly ever saw a car. Bought a truck for $1,100.

"Married in '52. I was thirty-two years old when I married."

"What did you do for entertainment?"

"'Grand Ole Opry' on battery radio. Never owned a watch...didn't need one...got up at light and went to bed at dark.

"Lot of young people today wouldn't do what I did."

It's fair to say that the present generation would no more agree to live on a non-electric Earth than it would on the surface of a barren moon or a waterless Mars. There are isolated cases of the rejection of electrical power, but the vast majority will stand toe to toe with all the private powers that be—the unified cry will always be: "Let there be light."

In the words of J.K. Smith: "It all comes down to common sense: observing, being up to date, staying on top of what's going on, being aware, anticipating problems down the road and trying to plan for them."

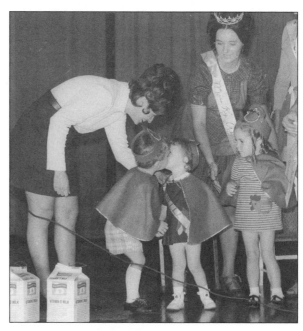

Annual co-op meetings are gatherings of cooperative participation, friendships old and just begun, and new beginnings of brighter tomorrows. (Big Sandy RECC archives)

BIG SANDY RURAL ELECTRIC
COOPERATIVE CORPORATION

Miles of Line:	1,020
Consumers billed:	13,197
Wholesale Power Supplier:	East Kentucky Power
Counties Served:	Breathitt, Floyd, Johnson, Knott, Lawrence, Magoffin, Martin, and Morgan

ADMINISTRATION

Silas Dean Wolfe	(dates unavailable)
Dan Gambill	(dates unavailable)
Bill Wells	1969 – 1995
Bruce A. Davis Jr.	1995 – 2003
Bobby Sexton	2003 – present

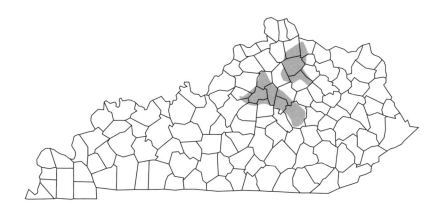

BLUE GRASS ENERGY
COOPERATIVE CORPORATION

We're not here to make a profit—
we're here to serve you.

Dan Brewer
President and CEO
Blue Grass Energy
Cooperative Corporation

Blue Grass Rural Electric Cooperative Corporation, named by J.L. Miller, a Madison County farm agent, received its first loan from the Rural Electrification Administration on October 1, 1937, in the amount of $120,000.

The first lights of the cooperative were turned on June 15, 1938, for a total of 258 connected members and 126 miles of energized lines. These lines served sections of Jessamine, Fayette, and Madison counties.

"I guess Momma was more satisfied with electricity than any of us. Momma got rid of the old coal oil lamp," Howard Teeter of Jessamine County remembers. "Daddy would follow us around when we turned the lights on and say, 'Helen, it's going to cost too much to run those,' and he'd flip them off."

In 1958 headquarters was built on Lexington Road in Nicholasville. In 1964, a Madison district office was constructed, and in 1991 a new Richmond office was opened at 2099 Berea Road. In 1995, a new headquarters office was opened, built on the lot adjacent to the old office.

In 1998, Blue Grass RECC merged with Fox Creek RECC in Lawrenceburg to form Blue Grass Energy, increasing the service area from six counties to fourteen. The Fox Creek District Office is at 1200 Versailles Road in Lawrenceburg.

When Blue Grass RECC consolidated with Harrison RECC of

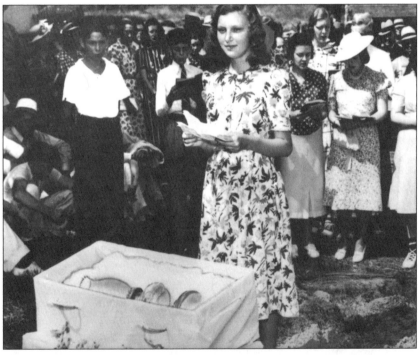

On June 15, 1938, the electric lines of Blue Grass RECC were energized to serve 258 members and part of the celebration was a funeral conducted for the kerosene lamp. Ms. Gloria Allender of Jessamine County "preached" the sermon while a chorus of Fayette and Jessamine 4-H members sang "Let The Lower Lights Be Burning" and "Brighten The Corner Where You Are." (Blue Grass Energy archives)

Cynthiana in 2002, the service area was increased from fourteen counties to twenty-three.

By 2002, Blue Grass Energy Cooperative Corporation had become a consolidation of Harrison RECC, Fox Creek RECC, and the original Blue Grass Energy. Today, Blue Grass Energy serves about 54,000 members in twenty-three counties, from Bracken County on the Ohio River to Madison and Estill counties.

In 2006, a new Cynthiana office opened at 327 Sea Biscuit Way.

"The first headquarters was at 113 South Main Street in Nicholasville. The property was leased from Mrs. Suzy Wagner at $25 per month. The fledgling organization had five board members. They were J.W. Moffett, Lexington, President; W.E. Overstreet, Nicholasville, Vice President; E.C. Moore, Secretary; E.L. Hersperger, Nicholasville, Treasurer; R.R. Davis, Nicholasville, Director; Phil R. Holloway, Nicholasville, Manager; and R.L. Bronaugh, Nicholasville, Attorney." (KAEC archives.)

"The first manager was Phil R. Holloway, Nicholasville. He was the first employee with Miss Ruby Clarke, bookkeeper.

"Forerunner to Blue Grass RECC was the Jessamine County Electrification Association. County Agent Grover Routt, Jessamine County Farm Bureau lawyer Robert L. Bronaugh and directors were early organizers.

"Electric substations were built at the Frank Scott Farm on Baker Lane and on the Hubert Ham Farm in the Poosey Ridge area of Western Madison County, just across the Kentucky River from Little Hickman.

"Transmission lines stretched between Baker Lane, 1½ miles north of Nicholasville, to Cottonburg in Madison, an estimated distance of twelve miles. Kentucky Utilities, which then served only Wilmore and spots along US 27, sold wholesale power to the cooperative.

"Nicholasville in those days generated its own electrical power from a coal fired boiler system.

"A lot of local help, horse teams, and wagons were utilized to make the power line sweep past Nicholasville through southeast Jessamine County to the Madison County connection.

"Records of the co-op reveal that Bradie Teater of Nicholasville was paid $2.50 on July 16, 1938, for hauling utility poles. Granville Logan accepted $6.00 for twelve ferries across the Kentucky River."

Howard Teater: "Back when I came back from the service, I worked for REA. We had one truck. The poles were one hundred and some feet tall. We had two poles and one bolted in the middle of it. I got to the top and was afraid to come back down. Luther Perkins had to come up and help me back down."

"About two thousand two hundred people attended the co-op's annual meeting Thursday, June 14, 2007, at Anderson-Dean Community Park in Harrodsburg.

"Members conducted official business, including the re-election of Danny Britt, District 2, Jody Hughes, District 6, and Brad Marshall, District 10, to the board of directors, and they heard reports from Blue Grass Energy officials.

"Our main message was one of solidarity. In this time of escalating costs and ever-tightening environmental standards, it is more important than ever that we stand together as not just a cooperative, but as a family.

"Blue Grass Energy's members are the greatest part of that family, and we will take these challenges and turn them into opportunities to better serve you—our member-owners." (Blue Grass Energy archives.)

"We continue to explore new ways to be more energy efficient and create clean, renewable energy. We will continue to offer programs and tips to help you save money and energy.

"Working together, and with cooperatives across the nation, Blue Grass Energy will continue to provide you with the lowest possible cost.

"And in the spirit of giving you the best service—service that extends beyond your electric lines—we are very pleased to award $10,000 in scholarships every year to graduating high school students in our twenty-three-county service area. This year, ten students received $1,000 each and were recognized at the annual meeting.

"The goal is to teach students the history of our country and the role electric cooperatives play in the national government."

Stories told by people before the arrival of rural electrification, whether the service would one day be provided by a private company or

Blue Grass safety and engineering employees gathered with consultants from the University of Kentucky in 1953. Some of those in attendance were Herman Brawner, Dr. Baker (UK), J. Gray, Ray Foster, Gus Harlow, Leroy Staples, J. Shehee, Stanley Taluskie, Bill Hinkle, Hobe Adams, J.R. Burnside, Walter Kaufman, Buck Jenkins, W.R. Short, Virgil Mitchell, Cecil Bush, Mr. Yeoman (UK), Joe Bradley, Curley Brown, Morgan Hill, Holley Warford, and Horace Hume. (Blue Grass Energy archives)

a newcomer cooperative, make plain the deepest of human need.

Chester Powell moved with his family to Todd's Road in Fayette County in the mid-'30s.

"The old farm house had stood empty for almost four years. Three members of the family had died there in the same year under tragic circumstances. The heartbroken widow and mother moved back to her home place and abandoned the house.

"The house was in sad repair from being left alone for so long to bear the wind and rain alone. The wavy glass windows would cast an eerie glare at the full moon rising over the tobacco fields or from the lights of a car passing on the road. But with some paint and paper, polished windows and the warm, wood burning cook stove, the place became my first installment of what I thought Heaven must be like.

"We did not have running water, no central anything really, and even though we lived only five miles from town we did not have electricity. Sewing, reading, and homework were all done by the light of kerosene lamps placed strategically around the room.

"When our fortunes permitted, my dad bought a nice RCA brand radio whose lustrous cabinet gave all the other furniture in the living room courage to look better. The radio was energized by a six-volt automobile-type battery. My dad was in charge of budgeting the amount of time we could listen to it before the battery would have to be taken to town and re-charged at a cost of fifteen cents.

"We listened to Lowell Thomas, 'Amos 'n Andy,' the morning news and farm report from Cincinnati, and with that item came the singing couple Lula-Belle and Scotty singing a hymn. The high point of the week was Will Rogers, all the way from California. He came on and went off with the sound of an alarm clock. The whole family laughed

Door prizes bring smiles at annual meetings, but the biggest prize of all is electric power in homes where once there was none. (Blue Grass Energy archives)

with him whether we really understood what he was talking about or not. I remember my mother asking my dad, 'Jim, what is he, what is his job?' He replied with feigned patience, 'Well for one thing he is a writer, and he is also a Democrat.' I didn't know a lot in those days but I knew for sure we were Democrats. One of Mom's friends told her about a program that came on about two o'clock in the afternoon called "Our Gal Sunday." Well, Mom ignored the budgeting rules and turned it on. She was hooked, and it was awhile before she confessed. The confession came as

On icy days when Mother Nature knocks out the electricity, there is no more welcome sight than a lineman tripping the transformer to turn the juice back on. (Blue Grass Energy archives)

supper was being put on the table and she was instantly forgiven for the violation. 'Mom,' he said, 'I was referring to the boys.'

"One more thing I recall was that we and at least two of our neighbors had cows to come fresh about the same time. We had more milk, butter, etc., than we could hardly manage at times. Our only means of cooling was a big ice box that held either fifty or seventy-five pounds of ice. It did a pretty good job if we didn't chip off too much for cold drinks, and we lost lots of milk to the hogs.

"One day my dad saw an ad in the *Lexington Leader* advertising a refrigerator that operated on kerosene. We went to town the next Saturday morning to check it out. The man who operated the store, which catered to farm folks, was very cordial. We asked about the ad and he said, 'I have one hooked up in the back and I'll show it to you.'

"To say the least, it was beautiful, snow-white with bold letters across the door spelling 'Kelvinater.' The owner told my brother and me to stand in front of the door, and he opened it. We would never forget that burst of cool air. He closed that door and opened the one at the top. He explained it was the freezer and he took out a paper box and handed each one of us a frozen fruit bar on a stick. Whereas before we had been hooked, we were now stricken. The gentleman's name was Charles Jett. Years later he was instrumental in starting the Bank of the Blue Grass.

"Mr. Jett turned to my dad and said, 'When do you want it delivered?'

"My dad had already seen the price tag hanging on the door handle, and he grinned and said, 'Now we sure do like it but there is no way we can afford it.'

"'Oh, I think we can work it out. Let's go into the office.'

"He opened the door to all of us and then closed it. He sat down at his desk and took out a sheet of paper with some numbers and lines on it. Then he turned to my father and said, 'Jim, my part is to trust you. You tell me when and where to deliver it. Your part will be to trust me. You pay me whatever you can whenever you can and if you agree, let's shake hands.' They shook hands, Mom dabbed her eyes, and my brother and I went outside and slapped each other around a little, which was our custom when we were happy.

"By four o'clock that afternoon that beautiful refrigerator was standing like a monument on our closed-in back porch.

"We were paying fifteen dollars a month rent on our house, thirty acres and all the barns and other buildings. It was either 1936 or 1937. I simply do not remember the year electricity began to slowly creep out our way. There was the digging of post holes, the setting of poles, and finally the stringing of wire. In the meantime, two men came and wired the house. We were ready but it seemed like an eternity before the wire was finally brought into our very own meter. Finally, the big day came.

"We had electricity!

"Simple as the system was, we had lights hanging down from every ceiling in every room of the house. We could play the radio all we wanted to, we could read as long as we wanted to, and slowly we were able to buy simple appliances that were not expensive. Mom had had to iron by putting flat irons on the cook stove. She

Teamwork is not the exclusive domain of linemen ... women participate as well. Blue Grass employee Stacy Adkins gets a helping hand and a lesson in pole climbing from a lineman during a safety exercise. (Blue Grass Energy archives)

would iron until one cooled down and replace it with a hot one.

"Then one Saturday we went to town and stopped by Mr. Jett's place of business. While we waited in the car, Dad went in and made the final payment on the refrigerator. He had sold a prized Jersey calf to a gentleman who lived near us so he could pay the debt. When he came out he was holding a box in which was an electric iron. He handed it to Mom and told her it was a gift from Mr. Jett. Her big brown eyes filled with tears that ran down her cheeks."

There was more than tears for Howard Teater: "My grandfather, Papa (Elvin) Reynolds, was so happy about (getting electricity) that he went out and bought himself an electric ice cream freezer, because he was tired of sitting there with that old hand crank, because every Sunday he made ice cream!

"When Momma Reynolds got electricity she wanted one of these Philco radios. She'd had that old battery radio and all she ever got was static. When she got this electric radio, she'd sit there and listen to WHAS and Renfro Valley every Saturday night."

BLUE GRASS ENERGY COOPERATIVE CORPORATION

Miles of Line:	4,517
Consumers billed:	54,465
Wholesale Power Supplier:	East Kentucky Power
Counties Served:	Anderson, Bourbon, Bracken, Estill, Fayette, Franklin, Garrard, Grant, Harrison, Henry, Jackson, Jessamine, Madison, Mercer, Nelson, Nicholas, Pendleton, Robertson, Scott, Shelby, Spencer, Washington, and Woodford.

ADMINISTRATION

Phil Halloway	1937 – 1948
Richard Forrest	1948 – 1951
Overton Giles	1951 – 1970
Peter McNeill	1970 – 1975
Jack Taylor	1975 – 1992
Dan Brewer	1992 – present

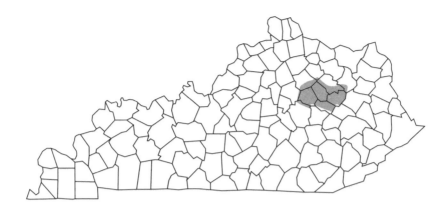

CLARK ENERGY
COOPERATIVE, INC.

The landscape will change, whether we like it or not,
the suburbs will continue to expand into the rural areas and
cooperatives will continue to grow.

Paul Embs
President and CEO
Clark Energy Cooperative, Inc.

Clark Energy, organized in 1938, at present serves 25,000 consumer-owners in Bath, Bourbon, Clark, Estill, Fayette, Madison, Menifee, Montgomery, Morgan, Powell, and Rowan counties. The cooperative, which changed its name to Clark Energy Cooperative in 1998, includes parts of rural Fayette, crossroads of Interstates 64 and 75, and the Bert T. Combs Mountain Parkway.

The first meeting of Clark Rural Electric Cooperative Corporation was held on Court Street in Winchester, March 17, 1938. The meeting was called to order by one of the incorporation members, J. Hughes Evans of North Middletown. He was chosen chairman of the meeting, and E. Ward May, also an incorporator and member of the corporation, was chosen secretary of the meeting. R.R. Craft was named attorney and

A.C. Lockridge was named project superintendent. In April 1939, Lockridge was dismissed and Evans was named acting superintendent. In November 1939, T.E. Steele was named project superintendent.

J. Hughes Evans lived on the family farm on Pretty Run Road in Clark County on the edge of Bourbon County. A member of the family, Mary Louis Evans recalls, "Before electric power came to the rural areas of the county, the house on Pretty Run ran electric off a Delco plant in the basement of the home. The basement had a concrete floor, rock walls, two windows, and a concrete stand for the Delco plant. The batteries were kept in clear or greenish glass containers approximately five and one-half by seven inches wide and ten inches deep. 'DELCO LIGHT' was on each side, with 'WATER LINE' marked at the top. 'KXG 13' was on the bottom. J. Hughes Evans gave me one of the containers and sold others for ten dollars. The containers made sturdy, miniature aquariums in future years after electricity arrived."

In most rural areas, many landlord homes had Delco units and

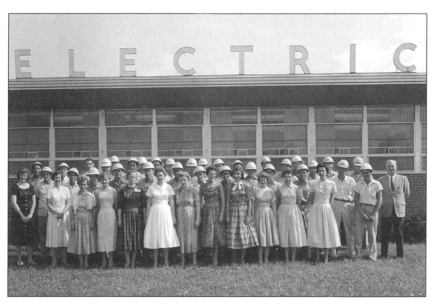

When J.K. Smith said "Let There Be Light" should be the story of people, he didn't mean just a few, he meant a community of Kentuckians. These are just a few who served Clark RECC in earlier days. (Clark Energy archives)

sometimes gas lights, but tenant houses typically relied on kerosene lanterns. Agricultural leaders such as J. Hughes Evans realized the disparity and dedicated their days to correcting it.

According to the Clark Energy archives, the Board of Directors in 1949 included: R.R. Craft, attorney; Dr. E.E. Curry, Secretary-Treasurer; J.L. Skinner, President; Walter S. Meng, Vice President; Manager William R. Hanshaw, John P. Greenwade, George Ginter, J.W. Deatherage, Beall Hadden, Edgar Rose, and Virgil Barnes.

Ms. Emma Hunt was the proud winner of a percolator that her daughter said, "...she used until she plumb wore it out!" Annual co-op meeting prizes are treasured incentives to "be there" to continue the cooperative tradition. (Clark Energy archives)

Dr. E.E. Curry's son, William, remembers how his father went door-to-door in 1938 trying to convince farmers to invest five dollars as a requirement to sign up for electricity. "He didn't have nearly as much success with the men as he did with the women in signing up. He'd talk to the man in the daytime and they'd say, 'No, we don't have the money.' But if he waited until nighttime, he'd always take an electric iron. Next time, in two or three days, there'd be the five dollars."

Dr. Curry's grandson, Everett, continues the Curry legacy of membership on the Clark Energy Board of Directors. "He (Grandfather

Curry) had an appreciation for electricity; he knew what it could do for people."

Mrs. William Curry is asked, "How much difference do you believe electricity has made in rural areas?"

Her reply: "Tremendous...an end to drudgery."

Clark County, an area richly diversified, is home not only to Clark Energy Cooperative but also to East Kentucky Power. Who decides who receives and who provides electric service in such a demographically diversified area?

"The state basically decides," says Clark Energy President and CEO Paul Embs in his office on the east side of Winchester. "In 1970, there was legislation that drew up territory lines—boundary lines. Those areas that are in dispute are handled by the Public Service Commission."

"What do you see as the value of the Kentucky Association of Electric Cooperatives?"

"One individual cooperative couldn't start to do all the services that KAEC does for us. All co-ops being part of the same organization gives us a bigger voice in Frankfort."

"*Kentucky Living* magazine?"

"It couldn't be done individually [by individual co-ops]. We couldn't afford to do it. We're not big enough to have a staff to do that all the time."

Payroll?

"Fifty-three full-time employees, but we contract for tree trimming and they have thirty-plus folks. We spend approximately a million dollars a year on trimming trees.

"Clark Energy Cooperative is a member of Touchstone Energy, which serves over 30 million customers in forty-six states. As a Touchstone Energy Cooperative we embrace the values of integrity, accountability, innovation, and commitment to community. We are committed to serving our customers. We are committed to serving our members."

Meet John Rainey, eighty-six years old, who lives on Thatcher's Mill Road in Bourbon County: "It was 1940 when electricity [Clark Energy]

came through—
stopped at our
house. There were
nine children, six
boys and two girls
(one boy died
young). We all had
jobs—some cut
firewood, some cut
kindling, some
milked cows, some
fed the hogs. We
had a routine. Had
to use horses and
mules. We had
carbide
light...didn't work
too bad. Now we'd
be lost without
electricity—lights
for stripping
tobacco, water
system one time
for five thousand
turkeys."

Many times the cooperative work is a single person measuring and charting the miles of rights-of-way. Jerry Rose stakes the new line serving the North Fork/Star Gap area of Powell County. (Clark Energy archives)

"How did you get water to the house before the lights went on?"

"Had a cistern at the back door. It was fixed up modern with a small pump inside, which had to be primed. Had a Warm Morning stove with a water reservoir. We took baths in a number 3 tub."

"Outhouse?"

"Three-holer."

A visit with Joe and Betty Jo Turner in their home on top of a hill close to the joining of the county lines of Bourbon, Bath, Montgomery, and Nicholas: retired school bus drivers, they look back to the days when there was no electricity in dark, rural areas.

"I was born in '33," says Betty Jo, who recently had a pacemaker implanted. "I knew when I went to Granny and Pappy's they didn't have electricity. As far as I remember, there was no stove upstairs. I believe some of the cooking was done there, or the big family cooked together and ate together in the kitchen. I remember my stepmother washing dishes in the room (upstairs), and she'd take the dishwater and throw it out the window.

"I remember the Aladdin Lamp on a big old library table, and it was sitting in what they called their sitting room. I'm sure both sister June and brother Johnny were in school. I remember them doing their homework in front of that lamp. They had other lamps, but they didn't put out the light like the Aladdin Lamp.

"The only heat we had was the fireplace, and I remember I had to go from that room down the long hall to the kitchen, and it was so dark, you know, during the early morning hours. We ate there at the table, and Granny was not in very good health.

"My mother had died when I was seven days old. There was no electricity and [the early morning] was like staring into the dark, more or less. It was so warm the day I was born, they had the windows up. But the day after I was born there was snow on the ground. My mother had a kidney failure, but we didn't go to the doctor back then—and she died of kidney poisoning.

"But Christmas time was a special time and there was always a tree at Granny and Pappy's, but there were no lights on it because there was no electricity. Granny made jam cake. And it would take hours, because they'd have to cook it in the old coal stove, and they had to mix it by hand, because they didn't have mixers."

"Joe, what difference did electricity make outside on the farm?"

"We had a refrigerator that was run by gas, and we took it with us wherever we went. It was easier when electricity came. When you had electricity you could pump water out of the creek and put it wherever you wanted it. You had power tools—if you had a drill it would be run by electricity. You did most of the repair work in the dark after working in the daytime."

"It was dark?"

"You had a lantern."

"What do you remember your mother and father saying about how it was before electricity came?"

"I don't remember them talking about it," says Joe. "They were just glad to have it."

"The coming of electricity seems almost like a dream to me," Betty Jo smiles a gentle smile.

Nancy Mason of Bourbon County remembers the days of cutting ice off the pond, using a team of horses or mules to bring the chunks to the icehouse, breaking up the pieces to fill the icebox in the kitchen—and make ice cream!

Anne F. Caudill, widow of Harry Caudill, author of *Night Comes to the Cumberlands*, remembers the days when electricity finally arrived in

Meetings, meetings, meetings. It takes a blend of clear thinking, unselfish intentions, and several prayers for inspiration. A former Board included (L-R, sitting): Jack Ginter, Jim Shultz, William Nelson Curry, Elmer Johnson, President & CEO, and (standing) Seldon Fannin, Paul Faulkner, William Shearer, Edmond Burgher, John Tabor, and Dan Yates, Cooperative Attorney. (Clark Energy archives)

Clark, Harrison, and Montgomery counties: "I was in college when REA reached my grandmother's family farm. Though the family lived in town from November through April or May and had since 1914, the whole clan returned to the old farm home for the planting, growing, and harvesting months. I spent many weeks and took as a matter of course the use of coal oil lamps. It was my job to clean and refill the lamps each day. I still have the little copper base of the small lamp I carried upstairs to my bedroom at night, and for forty years at Whitesburg I kept the old Aladdin lamp filled with oil at the ready."

Former Governor Bert T. Combs, swept away by the Red River in Powell County on December 3, 1991, was one of KAEC's Distinguished Rural Kentuckians. When he received the high honor, he recalled some earlier remarks:

"My mind takes me back to travel by mule-back over creek-bed

Sometimes the weather doesn't cooperate. Ice storms wait for no invitation. Downed power lines are challenges to rebuild and put back together again with teamwork. (Clark Energy archives)

roads; to open grate fires in winter and coal-heated cook stoves year-round; to coal oil lamps and outdoor privies. One had the feeling the men folk were not wholly convinced that book learning was here to stay. But their faithful wives, who spoke little in public but were usually opinionated and influential at home, had come down hard to build a new school. It would include the principles of the New Testament along with the subject matter of the lesson.

"My parents sent me to school at Oneida Institute when I was eleven. The year was 1922, and I was in the seventh grade."

With the passage of six decades, Governor Combs's widow, Sara, now Chief Judge of the Kentucky Court of Appeals, remembers the first day the lights came on at their newly constructed log house in Powell County.

"Electricity came to the nation's countrysides and valleys before we built Fern Hill in 1984. My husband recounted that for many people, the first sight of illuminated houses glowing in the night produced terror—fear of fire—rather than delight.

"Although not very remote in time from that first lighting of rural America, Fern Hill is geographically isolated in a valley nestled among the mountains with no neighbor in eyeshot. All the summer long of 1984 we watched the cabin rise log by log. We'd leave our bi-weekly inspections before dark, as there was nothing more to be done after the sun departed.

"And then one evening in the early fall, just a few weeks before construction was completed, we arrived later than usual for a quick visit before nightfall. Clark Rural Electric had worked its magic that very day! A glow filled the house and its warmth emanated from within, beckoning us up the driveway and inviting us to linger a while without the sun. Fern Hill at last had a heartbeat.

"No wonder life began in Genesis with the Lord's command, 'Let there be lights.'"

CLARK ENERGY COOPERATIVE, INC.

Miles of Line:	2,976.251
Consumers billed:	25,839
Wholesale Power Supplier:	East Kentucky Power
Counties Served:	Bath, Bourbon, Clark, Estill, Fayette, Madison, Menifee, Montgomery, Morgan, Powell, and Rowan

ADMINISTRATION

A.C. Lockridge	1938 – 1939
J. Hughes Evans	1939 – 1940
T.E. Steele	1940 – 1947
William Hanshaw	1947 – 1962
Cephas Allen	1962 – 1966
Elmer Johnson	1966 – 1986
Overt Carroll	1986 – 2005
Paul Embs	2005 – present

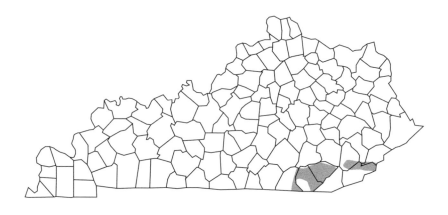

CUMBERLAND VALLEY
ELECTRIC

CVE plans to continue the provision of high-quality electric service for its members at the lowest possible cost.

Ted Hampton
Manager
Cumberland Valley Electric

The main office sits on the south side of Highway 25E between Corbin and Barbourville, where Pine Mountain stretches away to the east through Bell County and on to Harlan, Knox, Leslie, Letcher, McCreary, Whitley, and north to Laurel. It's the land of John Fox Jr.'s *Trail of the Lonesome Pine* and *The Little Shepherd of Kingdom Come*, home of Silas House and *The Coal Tattoo*, Gurney Norman's *Kinfolks*, Harry Caudill's *Night Comes to the Cumberlands*, and Harriette Arnow's *Seedtime on the Cumberland*.

Indeed, it was seedtime in the decade after President Roosevelt had prepared the soil for light where there'd been none before. The creation of rural electrification would come not without heartache and bruising controversy in the coalfields. The prime source of fossil fuel energy would mean the laying bare of idyllic mountains, and little creeks like Troublesome over in Knott County would tumble down with the inevitable side-effect stains of "progress."

Caudill's classic *Night Comes to the Cumberlands* could not push back the days when *light* also came to the

Mule power—the gee and haw of it, the up and down the hill and valley of it—brings light to distant far-away places. The mules know there might be an extra ear of corn in the trough after tough days of pulling wire. (KAEC archives)

Cumberlands. There would needs be a compromise. In 1940, eastern Kentucky coal would move by rail and by truck to East Kentucky Power, where it would be generated into electricity serving more than half of the Commonwealth. Restoration of the coalfields and the mountains to their very tops would become the challenge for another generation in a new century of converting fossil fuel into the lighting of homes downstream as far away as Lexington and Louisville.

The Kentucky counties served by Cumberland Valley Electric include some of the Commonwealth's most difficult topographies. The highest summit in the state, Black Mountain in Harlan County—4,145 feet above sea level—is surrounded by remote areas along streams—Gap Branch, Maggard Branch, Left Fork, Razor Fork, Deep Gap Branch— arising slender, unlikely home sites left in the lowering darkness until rural electrification finally reached the unlit headwaters of Poor Fork of Cumberland River. (A large part of eastern Kentucky is served by investor-owned utilities—American Electric Power [AEP] and Kentucky Utilities [KU].)

It was the challenge of Cumberland Valley Electric to illuminate the way to places where private companies made decisions not to venture. Other writers—Jim Wayne Miller (*The Mountains Have Come Closer*) and George Ella Lyon (children's books) have helped to bring another light where for so long up the deeply isolated passes there were mainly coal oil lamps, candles, Bibles, and *McGuffey Readers*.

A conversation with Ted Hampton, CEO since 1964 of Cumberland Valley Electric:

"We're very conservative—ten-year-old trucks, vehicles with 150,000 miles. We give good service. Pay our people well. Look for ways to cut costs. You've got to use your common sense, not a lot of

These three boys exhibit the faces of optimism, the holders of the future, the joy that comes with something as basic as a light bulb. Great fun and much cause for laughter holds forth at annual RECC meetings. (KAEC archives)

For these linemen in training in 1953, it took strong legs and well-placed hands and respect for hot juice to bring power to pole to transformer to home and barn. It still does. (KAEC archives)

frills. Always have been [conservative] since I've been here. That's why we have one of the cheapest rates of the cooperatives. We're constantly looking for ways to cut costs. Meet with the staff every Monday morning, and we try to look forward to see how we can help our service, but if we don't, we look back to see what we *could* have done.

"We spend $800,000 a year—one of our biggest expenses—for tree trimming.

"Outages are automatically read at the office. Our co-op is the most mountainous—nine counties, two in Tennessee—all rock—have to blast holes in the ground. 'Dozers make roads for the trucks to use to set the poles. Being most mountainous, the terrain is the number-one problem. And it's all rock—you've got to blast your holes in the ground, and then you set your poles. A lot of companies have trucks, but we use 'dozers to set the poles. In several cases, we have used helicopters to help set the poles.

"I've been forty-two years with Cumberland, started when I was twenty-two in 1964. Cumberland Valley was formed and in 1942 was when lines were started to be run. Bill Hampton, my uncle, was CEO from 1954 to 1964. I've been CEO from 1964 to the present day. There have been only five: Jimmy Broaddus, W.N. Jackson, Marion Thacker, Bill Hampton, and me.

"We have fifty-two employees.

"I was a school teacher prior to coming here. I graduated from Cumberland College [now University of the Cumberlands]. I taught PE and history. I graduated in 1962 and started working immediately upon graduation.

"I attended Little Brush Creek school. Went through eight grades. No electricity. That came later, but when I started, there was none, not until 1948. We had a big old pot-bellied stove right in the middle of the school. Never did have air conditioning in it, no electricity. For heat, we just had the coal stove. There were twenty to thirty students of all ages— one teacher."

In 2007, Mark Abner, Cumberland Valley Electric's engineering manager, listed what he believed are CVE's three most significant achievements:

1.) CVE's outside plant was completely reconstructed approximately thirty-five years ago.
2.) Beginning in 1997, CVE began deployment of Hunt Technologies' TS1 Automated Meter Reading infrastructure. That system is gradually being replaced with Hunt's TS2 system. The conversion should be complete by about 2013.
3.) CVE has, since 2004, successfully implemented an electronic Geographic Information System (GIS) to replace paper-based mapping that had been used since the inception of the cooperative.

A conversation with seventy-seven-year-old retired coal mine foreman Billy Cain, thirty years a strip miner—words that have become as volatile as pork fat on a hot skillet:

"Tell me about reclamation of strip-mined land, Billy."

"People are building homes, airports on stripped flat lands," says he, the issue of mountaintop removal seeming problematic, never mind how well-intentioned. Kentucky's strip mine law was enacted in 1966. In 1981, Kentucky led the way to national enforcement of strip mining legislation. In the twenty-first century, the search for fossil fuel has led to a new flash point called "mountaintop removal." The deepening irony is that relatively cheap coal, coveted and extracted, was responsible for the generation of electricity, and it in turn had made life in the Cumberland Valley vastly different. For that, Billy Cain would be grateful throughout his long life.

"My mother and father didn't have electricity. We went to bed a little after dark and got up before daylight. Some went to the barn to feed stock, some started breakfast. There was no refrigeration. We had coal oil lamps.

"My neighbors didn't have electric. Just coal oil lamps. My aunts and uncles didn't have electricity—they had coal oil lights. We had to draw water out of a well—on wash days, if the rain barrel was empty, I had to pack it from a sawmill pond, 500 yards away. There was no such thing as city water—we didn't get city water until 1955."

Billy remembers, too, those long-ago days of Little Brush Creek school—that one room with one teacher and eight grades... "no electricity, just a potbelly stove for winter, but no air conditioning" for the red-hot days of summer. "Nobody had running water. Washing clothes—if it hadn't rained, I'd take a lard can and a little red wagon to the sawmill pond."

When the lights finally did come on, "Neighbors burned it all night. They sat up and looked at it. It was really something. The family below me had thirteen kids in it and they didn't get electricity until up in the '50s. Doors weren't locked.

"Married early, growed up quick. No big upty-dumpty.

"My mother was one strong woman. She loved to cook. We had two grates and a cook stove. We washed in a number 3 washing tub. We had a number 3 in the kitchen and with the oven door open, we washed off— all kids went barefooted. [Fodder for Sunday morning comic strips depicting Kentucky "hillbillies,"—L'il Abners and Daisy Maes, unshod,

The Brush Creek one-room school, where once there were no lights, no air conditioning, no electric heaters, just good hearts, willing minds, and strong-willed people. (Cumberland Valley Electric archives)

unbright, and curiously funny—never mind true cause and effect.]

"We had a Jersey cow. I had to milk her, but I can't remember how young I was. In the summertime I had to walk her a good mile to pasture, then I had to get her in the evening. I had to get in coal kindling. If you had milk, butter, and eggs, you could make it [in life].

"There was an ice man. We put twenty-five pounds in our ice box with the butter, eggs, and milk—lots of rich milk. Mother raised a flower garden. If there was a sick family, she'd cook for them. Took care of each other. If there was a funeral in the community, my mother cooked. The families kept up two nights. My father would help dig the grave, but he didn't go to the funeral."

Life changed for Billy Cain.

"Mother was bedridden for about a year. One night, it was my time to get up with her, I heard her and put my feet on the floor and heard her singing *Amazing Grace*. Then she started praying, talking to the Lord, and she said, 'Jesus, what's going to happen to my little boy?' and then she said, 'Jesus, haven't we suffered enough? When are you going to come get me?' By then, I had tears dropping to the floor. Two months after that, she was gone.

"She was forty-six when she died of cancer. We were awful close. I was fourteen when Mother died.

"Father died four months later."

Billy Cain's voice tightens. There are tears.

"My daddy was a big man. Stout. Didn't know his strength. But he had a tender heart. He was working in the coal mines—cross collars, four inches thick, eight inches wide, eighteen inches long. Busted the big leader from his heart. It was March of '45. By the time they got back from fetching help, he was gone."

Billy also worked in the mines when he was a youth—Bell and Leslie counties, "Mighty rough work—come awful close a time or two of

These are seven of the many grassroots workers, those in the beginning, the farmers as linemen ... fourteen hands, fourteen legs, but one calling, "Let there be light" ... for farms, too. (KAEC archives)

getting killed. I went in the big mines with my daddy. I got $4.05 for going in and I got sixty-six cents a ton for loading. That $4.05 was travel pay.

"I stayed with my aunt, and we didn't have electricity. Country boys and girls grew up quick. We'd take three trips to fill the coal bucket up, all you could pack. My aunt and uncle, they got natural gas before electricity. It came through their property so they got a natural gas Frigidaire. As for ice, I remember Daddy had come in with a big four-inch concrete tile, dug a hole in the corner, put the tile in the ground, built the floor. First he put rock, anything he could get down there. There was an ice box, but it only held twenty-five pounds. We'd have the ice man leave about seventy-five pounds—cooled the butter, eggs, and milk.

"Washing dishes? They'd have that water scalding. My sister would wash and I'd dry. Fried potatoes, fried hoecakes, and my mother could fry them where they were crisp all the way through. There was no air conditioning and we had four rooms—two bedrooms—my father and I slept in one room and my mother and sister in another.

"We had a Jersey cow—I've still got her butter mold. We traded out buttermilk and the eggs. Little girl come in, 'Mr. Billy, my mammy's said to tell you she's craving buttermilk.' When I grew up if anyone in the neighborhood needed anything, they got it.

"There was a battery radio, and we didn't hardly turn it on until Saturday night—'Grand Ole Opry,' 'Amos 'n Andy.' There'd be thirteen to fifteen people sittin' around it.

"When we got electricity my neighbors set up and looked at it. It was really something. It was way up in the '50s when they first got electricity. They all lived up behind Fightin' Creek and that was the last to get electric in the area."

"World War II?"

"Boys to Army, girls to Detroit. Dodge and Plymouth bodies," says Billy. "Worked nineteen hours a day for three months—seventy-seven dollars a week."

"And when the lights came on after the war? in the '40s?"

"Neighbors burned it all night. They sat up and looked at it. It was really something—a miracle."

Billy doesn't think Al Capp would understand.

CUMBERLAND VALLEY ELECTRIC

Miles of Line:	2,568
Consumers billed:	23,390
Wholesale Power Supplier:	East Kentucky Power
Counties Served:	Bell, Harlan, Knox, Laurel, Leslie, Letcher, McCreary, and Whitley in Kentucky; Claiborne in Tennessee.

ADMINISTRATION

J.S. Broaddus	1940 – 1944
W.N. Jackson	1944 – 1947
Marion Thacker	1947 – 1954
William Hampton	1954 – 1964
Ted Hampton	1964 – present

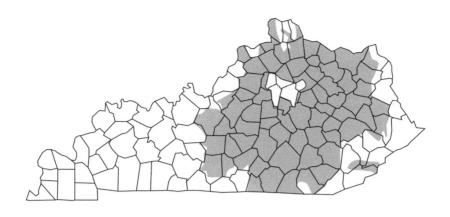

EAST KENTUCKY POWER
COOPERATIVE

East Kentucky Power Cooperative
was created by our members to give them a
viable competitive alternative for their power needs,
and we need to be focused every day on fulfilling that mission.

Robert M. Marshall
President & CEO
East Kentucky Power Cooperative

East Kentucky Power Cooperative [EKPC] was organized in the summer of 1941 but did not produce its first power until 1954. The delay was due to World War II and long legal battles with commercial power companies. After incorporation on July 7, 1941, EKPC's Board, on July 14, elected J.V. Swaim as its first President. He served until November 5, 1941, when William C. Dale was elected President. Dale served until 1950.

The underlying principle at EKPC is a commonality: people, consumers, and owners of their electric distribution cooperatives, having an inalienable right to operate and control their own *generation* facilities. They are beholden to themselves within the sanctioning shield of the state Public Service Commission.

EKPC, a nonprofit cooperative, generates—manufactures—its own wholesale electricity and supplies it to the owners, sixteen cooperatives in the eastern two-thirds of the Commonwealth. EKPC is one of the largest producers of renewable power in the southeastern United States.

Each of its member systems is represented on the co-op's Board of Directors. EKPC and its member systems plan and work closely together efficiently and economically to serve the needs of their consumer-owners.

EKPC opened its first offices on the fifth floor of the McEldowney Building in downtown Winchester. The year was 1951, the time of the Korean War and a winter blizzard. There were only seven employees. Among them were Manley Combs (who would spend the next thirty-five years with the cooperative), Christine Shelton Combs, and H.L. Spurlock, the first General Manager.

By 1957, EKPC had moved into its new office building at 4775 Lexington Road on the outskirts of Winchester, and by then there were nearly 100 employees.

Office staffs begin small and grow with time and necessity. Among the first few employees of East Kentucky Power were (L-R) Manley Combs, Christine Shelton Combs, Hugh Spurlock, Manager, Lois Bush, and Frank Belzer, Stanley Consultants contractor. (EKPC archives)

The decision to establish East Kentucky Power was based on the staunch belief held by thirteen distribution cooperatives that they would be better served by owning their own generation and transmission cooperative. The thirteen were: Blue Grass, Cumberland Valley, Fleming-Mason, Fox Creek, Harrison, Inter-County, Jackson County, Nolin, Owen County, Salt River, Shelby, South Kentucky, and Taylor County.

The Dale Power Station, named for William C. Dale, EKPC's first Board Chairman, is one of the oldest fossil-fuel plants in the national co-op network. Since December 1, 1954, at Ford, Kentucky. Today, the plant produces up to 196 net megawatts of power 365 days a year, 24 hours a day. (One megawatt [MW] equals one million watts.) (EKPC archives)

Today, East Kentucky Power Cooperative generates and distributes power to sixteen member co-ops:

Big Sandy RECC	Jackson Energy Cooperative
Blue Grass Energy Cooperative	Licking Valley RECC
Clark Energy Cooperative	Nolin RECC
Cumberland Valley Electric	Owen Electric Cooperative
Farmers RECC	Salt River Electric
Fleming-Mason Energy	Shelby Energy Cooperative
Grayson RECC	South Kentucky RECC
Inter-County Energy Cooperative	Taylor County RECC

William C. Dale Station is one of three coal-fired generating plants owned by EKPC. Located on the Kentucky River near Ford in Clark

County, Dale Station's facilities include four units: two 23,000 (net) capacity kilowatt units built in 1954; a third unit completed in 1957 rated at 75,000 (net) kilowatts; and a fourth unit in 1960 that is also 75,000 kilowatts.

On July 7, 1941, thirteen Kentucky co-ops met to formulate plans for the construction of a power station that they would jointly own as a part of the EKPC. They hoped this power supplier would ease the financial strain caused by the necessity of purchasing their power from various private utilities across the state. In addition to supplying them with cheaper power, the plant would also assure them of an adequate supply to meet increased power demands in the future.

On October 24, 1941, the first loan contract was executed with the federal Rural Electrification Administration. The attack on Pearl Harbor in December 1941 necessitated the voluntary suspension of EKPC's plans for the duration of the war. On December 14, 1950, after two months of bitter legal battles with private power interests, EKPC was awarded a certificate of necessity by the Public Service Commission. Activities again got under way. A plant site was selected at Ford; Hugh Spurlock, Jackson County RECC Manager, was named General Manager of EKPC; Clark, Farmers, Licking Valley, and Meade joined in 1949, followed by Big Sandy and Grayson in 1951. Eventually, Meade opted out of the system.

Groundbreaking ceremonies were held on November 23, 1951, and almost three years later, on June 12, 1954, the newly completed $12 million Dale Station transmitted its power to member cooperatives for the first time. An official dedication was held on August 28, 1954.

The William C. Dale plant was named after a man who was one of the leaders in Kentucky's rural electrification movement. Bill Dale's active and untiring efforts were instrumental in the pioneering, planning, and completion of EKPC's first plant. He was the Manager of the Shelby Co-op at Shelbyville from its inception in 1937 until his death in 1950. Bill Dale died while actively engaged in the battle to build the power station that bears his name. He suffered a heart attack while waiting to speak at the annual meeting of the Shelby RECC in August 1950 and died shortly afterward.

Two longtime friends of rural electrification in Kentucky, the distinguished late U.S. Senator John Sherman Cooper and his wife, Lorraine, stand in front of the Cooper Power Station that was named in his honor. First in operation in 1965, the plant, located near Somerset, Kentucky, today generates 341 net MW of power, enough electricity to power homes in thirty-one cities the size of Somerset. (EKPC archives)

Here are additional facts from the EKPC archives:

"The East Kentucky Rural Electric Cooperative had 1,070 miles of 69,000-volt transmission lines serving eighty-seven substations carrying electric power to approximately 137,000 rural farms, homes, schools, churches and rural industries of the 16-member cooperatives.

"Starting out with two 20,000 kilowatt generators in 1954, the East Kentucky Dale Power Station added a 66,000 kilowatt generator on October 1, 1957, to give the power station a capacity of 106,000 kilowatts. On December 30, 1957, the Kentucky Public Service Commission approved East Kentucky's application for permission to construct a fourth generator unit of 66,000 KW at Dale Station to meet the increasing power needs of the member co-ops. Without this unit, East Kentucky would have been unable to meet the total power requirements of its member cooperatives. These units were later upgraded to add generating capacity.

"East Kentucky now provides service to more than a million Kentuckians in the eastern two-thirds of the state.

"East Kentucky Power's member-distribution cooperatives have the responsibility of serving electric power to people in their service areas in the most economical and efficient manner possible.

"The Dale Power Station has a coal conveyor system capable of delivering more than 80 tons of coal hourly. In a twenty-four-hour period, the plant can use up to 2,000 tons of coal. The coal is used to make steam, which at a pressure of 1,250 pounds and at a temperature of 950 degrees, operates steam turbines, spinning at 3,600 revolutions per minute, which cause the generators to manufacture 13,800 volts of electricity. This electricity goes through a power transformer, which steps up the voltage to 69,000 volts. This power is sent out over the EKPC transmission system to rural electric cooperative members in eighty-seven counties. In 1958, the power plant consumed 183,852 tons of Eastern Kentucky coal and generated a total of 430,767,000 kWh to member co-ops.

"Most of the baseload electricity generated by Kentucky-based cooperatives comes from coal, excluding 170 megawatts of hydro peaking capacity received from the Southeastern Power Administration (SEPA) and fifteen megawatts from EKPC's landfill gas plants.

"Arriving at Dale Station by truck, the bituminous product has been delivered from surface and deep-mining coal operations.

Located on the Kentucky River at Trapp, Kentucky, and named for J.K. Smith, the Kentucky and national rural electrification program pioneer, the seven-turbine, combustion peaking plant operates on fuel oil or natural gas. The station, designed to run during peak times in summers (626 net MW) and winters (842 net MW), began operation in 1988. (EKPC archives)

"When delivered at the plant, the coal is placed in a large stockpile where it is continually being spread out and packed to prevent possible spontaneous combustion.

"Coal is transported from the stockpile to the plant by a conveyor system. Inside, it is then laced in bunkers and fed into pulverizers, which grind it to a consistency as fine as face powder. From there, air from fans picks up the tiny coal particles and blows them into a boiler firebox, where temperatures reach 2,500 degrees Fahrenheit. From here, the stages in producing electricity are set in motion.

"Very simply, coal is burned to boil water and produce steam that drives a steam turbine. The steam flow is directed against the blades of the turbine whose shaft turns a generator to produce electricity.

"The turbine-generators at Dale Station have a combined unit capacity of 196,000 (net) kilowatts.

"The turbine is essentially a windmill although far more complex. It has hundreds of blades, some which are stationary, others that rotate. These blades are arranged in groups called stages. Steam is forced into one stage and from that stage into the next, like a series of windmills with the same wind turning all the blades. The shafts of the turbine and generator are directly connected, thus the rotation of the turbine shaft turns the generator that produces the electricity that flows out through conductors.

"Electricity is then transmitted to a large power transformer where it is stepped up to transmission line voltage. After leaving the power transformer, the electricity goes to a switchyard and is distributed through the high voltage transmission system to the sixteen member distribution cooperatives.

"All of the water needed for Dale Station, whether for steam generation, cooling or drinking, is obtained from the Kentucky River. An intake structure, located on the banks of the river, makes it possible to remove the water, circulate it through the station and return it to its original source.

"Purified water is pumped into the system where it is preheated and moved into the boiler where it is changed to steam. The steam is superheated in the boiler to 950° prior to its journey to the turbine.

"When the steam has completed its travel through the turbine, condensers convert it back to liquid so that it can be pumped to the boiler for reuse.

"The condenser water, used to cool the boiler steam, is then returned to the river with only a nominal temperature increase.

"Through the years, East Kentucky Power has established a strong record of effort to preserve a clean environment for future generations."

Says CEO Bob Marshall, "EKPC has always been and continues today to be very sensitive and supportive of the environment. We are involved in a lot of environmental education programs. We were the first generator in the state to provide green power through our landfill power plants, and many other initiatives.

"We're dealing with a new generation of people. Twenty-five percent of the population today is under twenty-five years of age. Seventy-eight percent of our population does not remember when there was no rural electric program; therefore, we must gain their interest, not by talking about the pioneering benefits of the program but by what we're doing today for the good of all the people.

"East Kentucky Power Cooperative's members market the clean, renewable energy to members under the EnviroWatts brand name.

"EnviroWatts is offered to residents of Kentucky through fifteen of East Kentucky Power Cooperative's sixteen member systems. A voluntary program, EnviroWatts adds an additional $2.75 per 100 kWh [kilowatt-hours] per month for one year. Purchasing one block of EnviroWatts each month for one year has the same environmental benefit as taking the family car off the road for three months, or planting about 1 ½ acres of trees."

Rural Kentucky has graduated from the coal-oil lamp era and joined twenty-first century America. The generation and transmission cooperative came into being out of sheer necessity. Had it not been created, it is doubtful that the cooperative power program in Kentucky could have survived, much less grown to its present level.

The individual plants operated by East Kentucky Power are:

Dale Station (in Clark County)— Four units generating 196 net megawatts.

Spurlock Station (in Mason County)—Three units producing 1,118 megawatts: Unit #1, Unit #2, and the Gilbert clean coal unit. Another clean coal unit will be added in April 2009.

Spurlock Power Station, located on the Ohio River at Maysville, Kentucky, is the largest plant owned by East Kentucky Power Cooperative. Named for rural electric pioneer Hugh L. Spurlock, the three-unit, coal-fired facility produces 1,118 net MW of power. (One MW powers about 550 homes.) (EKPC archives)

J.K. Smith Station (in Clark County)—Seven combustion turbines producing 626 net megawatts in summer and 842 net megawatts in winter. Two more combustion turbine units will be added in 2009. Smith Unit #1, clean coal unit is scheduled to be added in 2012-2013.

Cooper Station (in Pulaski County)—Two units generating 341 net megawatts.

Alex B. Veech, past President of EKPC, captured the spirit of *Let There Be Light* in a copyrighted company publication in 1967.

"The younger generation cannot imagine what a thrill it was to see light in a home that had only known darkness at night; it was as if a new world had been discovered.

"Electricity, however, has meant more than just lights to the rural people of Kentucky. It has brought an end to many of the previously manual chores that made farm life a drudgery. Electric power also freed the housewife from back-breaking work that robbed her of strength and health. Electricity permitted her to spend more time with her family.

"Wholly owned by the distribution cooperatives, East Kentucky RECC has been instrumental in bringing rural Kentucky an abundance of economical wholesale electric power.

"This is a story of democracy in action in the best American tradition.... principles upon which this nation rests: perseverance and personal sacrifice for a good cause...strength through unity and government assistance in creating a better, fuller way of life for a large segment of the population that had been ignored by commercial enterprise."

EAST KENTUCKY POWER COOPERATIVE
ADMINISTRATION

H.L. Spurlock	1951 – 1975
Ronald Rainson	1975 – 1979
J.K. Smith (*)	1979 – 1980
Donald Norris	1980 – 1994
Roy Palk	1994 – 2006
Robert Marshall	2007 – present

(*) Interim President & General Manager

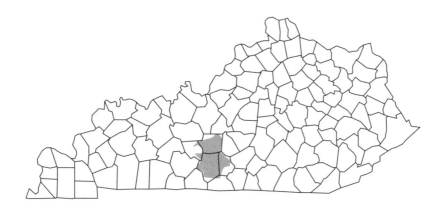

FARMERS RURAL ELECTRIC
COOPERATIVE CORPORATION

We never forget who we work for or why
we are in business.

William T. Prather
President and CEO
Farmers RECC

The first fifty years of history of Farmers Rural Electric Cooperative
Corporation is a testament to committed community service, a rich
tradition still alive today in the home office at 504 South Broadway in
Glasgow.

A promising view through a window of time was captured in *Let
There Be Light—A History of Farmers Rural Electric Cooperative,
1938-1988,* which the leadership published in 1989, then passed along to
future generations. And such is the intent of this current book, *Let There
Be Light*, fundamental to the formation of Touchstone Energy—a
national alliance of local, member-owned electric cooperatives doing the
people's business with integrity, accountability, innovation, and
commitment to community.

Strong leadership strung the lines along the pathway to fulfillment.

"March 15, 1938 was a day for which hundreds of South-Central Kentuckians had been waiting. On that date Farmers RECC was turned from a dream into a reality when incorporation papers were issued by the Commonwealth of Kentucky.

"During 1935 and 1936, REA field representatives surveyed Kentucky and their findings indicated rural Kentuckians were not ready for electricity and that electrification projects were not feasible in Kentucky.

"J.O. Horning, President of the Kentucky County Agents Association, did not accept the REA findings because he knew rural Kentuckians wanted electricity.

"Horning called meetings, and proposed electrical projects were drafted and sent to Washington, D.C. Still REA was unconvinced and it took persistent effort to convince REA that a second look at Kentucky was needed. Even after another visit by a representative and a meeting with Horning and other interested parties in Glasgow in February of 1936, REA reiterated that Kentucky was not ready for electricity.

"The matter did not end there for those at the meeting in Glasgow, and plans were made to prove to REA that Kentucky was ready for electricity.

"Immediately after Congress passed the Rural Electrification Act of 1936, and it became effective, county agents in Barren, Hart, and Metcalfe counties called meetings and organized survey teams to determine if rural residents were interested in receiving electricity through a cooperative of their own. After two weeks, a total of 1,275 people had expressed a desire for electricity.

"Meanwhile, the General Assembly of Kentucky had passed what was known as the Rural Electrification Cooperative Corporation Act permitting the operation of cooperatives in Kentucky.

"Maps made using survey information were forwarded to Washington, D.C. and in February 1938, a meeting was held in the Barren County Courthouse with REA representatives and it was determined it was feasible to set up one electric project for Barren, Hart, and Metcalfe counties as a joint effort.

"A committee of seven was selected—two from Hart, two from Metcalfe and three from Barren—to act for the proposed project. The articles of incorporation were filed for forming the Farmers Rural Electric Cooperative Corporation and a charter was granted on March 16, 1938.

"March 18, 1938, the first Annual Meeting of members was held and a Board of Directors was elected—Rodney Young, E.B. Hatchett and L.L. Wells of Barren County; C.L. Adair and H.C. Moss of Hart County; and F.E. Asbury and Joel R. Depp of Metcalfe County.

"Several months of hard work followed and on January 12, 1939, the switch was closed at the Goodnight substation north of Glasgow, energizing the first electric line for Farmers RECC and providing electrical power to the first 107 members along 51.1 miles of distribution line. This was the beginning of dependable electricity at low cost to consumers of Farmers Rural Electric.

"In November of 1947, the Munfordville substation was energized in order to relieve the Goodnight substation of some of the load.

The faces of the '50s portray a conviction that a rural electric board of directors should include a no-nonsense agenda for decisions by the people, for the people. (Farmers RECC archives)

Behind the meters, the poles, and the transformers were those who kept the required records–J.B .Galloway, Mary Porter Hardcastle, Marie Vance, Jean Kemp, Charlotte Bryant, Elizabeth Knipp, Ruby Pearl Shipley, Alton Porter, Frances Steenbergen, Garnett Vance, and Helen Joyce Redman. (Farmers RECC archives)

"On August 9, 1949, the Cooperative adopted the Capital Credits Plan. Under this plan, all funds paid in by members in excess of cost of operation were credited to the accounts of each individual member to be refunded annually to each member in proportion to his use of power.

"The first Miss Farmers RECC was crowned in 1950 at the first annual meeting held at Cavalry Field in Glasgow. She was Miss Wanda Matthews, daughter of Mr. and Mrs. Paul Matthews, Route 4, Glasgow.

"The third substation was energized at Temple Hill on August 31, 1951.

"The headquarters building on South Broadway in Glasgow was built and dedicated on June 4, 1954.

"The fourth substation was added to the system in August of 1954 at Knob Lick; the Beckton substation was energized October 15, 1959; the Cave City substation on October 1, 1971; and the seventh substation, the Parkway substation, on November 13, 1987.

"November 16, 1987, Capital Credit refund checks totaling $154,658.03 were issued to Farmers' consumers who received service during the years of 1950-1956.

"Farmers has grown from 107 consumers on 51.1 miles in 1939 to nearly 15,500 on 3,000 miles of line in 1989. (As of 2006, the numbers had increased to 23,537 consumers on 3,481 miles of line.)" Farmers Manager, J.B. Galloway, described the miles of line as made of "wire, steel, wood and the hard work of our employees."

J.B. Galloway was the fourth General Manager of Farmers Rural Electric and served in that capacity from August 1948 until November 1, 1988. He began his career at the cooperative in December 1945 as the electrification advisor until his appointment as General Manager in 1948.

Throughout his career, Galloway was a staunch supporter of rural electrification and he brought both state and national recognition to the cooperative.

Galloway often said, "Anything that makes life better, that makes work easier, that makes home life more comfortable and secure, improves the quality of life—that's the business we're in—not just selling electricity—but providing a service that in one way or another helps make life better."

Current Flashes, published by Farmers Rural Electric Co-op in October 1949, provides one example:

"For this month's visit 'Down on the Farm', we chose the farmstead of Mitchell and Irene Brents. Their one hundred and two-acre farm is located at Coral Hill in Barren County.

"Mitchell's principal crops are corn and tobacco and, of course, hay crops. The Brents have no children to help with the farm work.

"The most important use of electricity on their farm is located in the milk shed in the form of an electric milker. Mitchell thinks this is the most important piece of equipment he has on the farm. It is easy to understand why he feels this way, because he always averages about sixteen cows in this milking herd. He has had the electric milkers five years and says they more than paid for themselves the first year. Mitchell says the last five years have been quite different to the years before he got his electric milkers. Back then he milked ten cows by hand and his days were a lot longer and harder.

"We noticed a radio in the milk shed and commented upon it.

Mitchell explained that he believes the radio in his milk shed has increased his milk production. There is one thing about it that he is sure of, however, and that is it saves feed. He says in the past he fed the cows the entire time he was milking them, but now the soft music seems to keep them calm and contented, and he does not have to feed them until he is ready to milk.

"Mitchell stated that the next step he would take in modernizing his farmstead, after completing his home, would be to install an electric water system so he can have fresh running water when and where he needs it.

"In their home they have a refrigerator, washer, iron, radio and electric mixer. They also have an electric brooder which is of five hundred chick capacity. Mrs. Brents teaches at Oak Ridge school and of course, her time at home is limited, and she must make every minute count. With ELECTRICITY she is able to do this."

Modern decades (1985-2008) of the *Let There Be Light* story have reconfirmed Farmers RECC's expanding mission: "To raise the quality and convenience of its service to meet greater expectations of another generation of membership, and to continue to contribute to improving the quality of life in its service area."

During the late 1980s, the cooperative took a leadership role in working with Barren County local government to establish the IDEA industrial park to encourage and facilitate local economic development. The park has been very successful in attracting industry and giving a boost to the local economy.

In 1994, a devastating ice storm hit on February 10. Dedicated employees worked around the clock for approximately two weeks before service was restored. Approximately 300 poles had to be replaced. It was an extremely stressful time for both the membership and the cooperative employees.

Jackie Browning, the co-op's President and Chief Executive Officer (1988-2007), died May 4, 2007, from an apparent heart attack. His colleagues paid tribute: "Jackie ran the Farmers Rural Electric Co-op

No dwelling or structure, no occasion however modest, shall go unlit. Wherever there's a need and a desire there will be a pole, a line, a meter, and a workman. (Farmers RECC archives)

office for nineteen years in a style as steady, sure, and understated as he was in other parts of his life. He spent a total of thirty-five years at Farmers, starting in the engineering and operations department.

"Jackie kept the co-op at the innovative edge of the utility industry, expanding into the most modern business methods for serving the customer-members.

"Jackie's most notable feature was his commitment and integrity as a presence in our community. He was an especially dedicated member of Calvary Baptist church, where he served as a trustee, deacon, and taught Sunday school." (*Team Energy*, Farmers Rural Electric Cooperative 2006 Annual Report)

Continuity of management would be unbroken. William T. Prather is the new President and CEO of Farmers Rural Electric, his handshake firm and friendly.

"Progress continues with the cooperative recently completing the deployment of an automated meter-reading system in early 2008," he says. "Members no longer have to read their meters each month — something they have done since the cooperative began operations in

1938. Additionally, to improve efficiency, weekly cycle billing was introduced to more evenly distribute work throughout the month. Green power electricity produced from renewable sources was offered to the membership beginning in April 2008."

And the people? Some still remember very well how it was before the lights came on and, more important, what it means today. They take the time to write.

Polly Smith Spillman was born

Substations now dot the countryside as integral parts of the co-op's lines of distribution. On January 12, 1939, the switch was closed at the Goodnight substation north of Glasgow, energizing the first electric line for Farmers RECC. (Farmers RECC archives)

in Owl Springs, Kentucky: "This little space of heaven on earth is located in Barren County. My father purchased the one-room school house and converted it into a machine shop. Owl Springs was named because of hoot owls and some of the best spring water in the county.

"The hand pump at the house was on the back porch, which I thought was convenient. We would have to take turns pumping when Mama would do laundry. It took lots of buckets to fill the big washtub for clothes and washing kids.

"Most nights we were in bed early. But on certain nights of the week the neighbors would arrive to listen to 'The Lone Ranger' and other shows on the radio. Daddy had a generator in the smokehouse and had wired the house so we could have this occasional luxury.

"Then one day big trucks came and unloaded huge, tall poles that could touch the sky. As the linemen installed the power lines, they would come to the shop at dinnertime. They would look hot, thirsty, and tired upon arrival but when it came time to go, each would have a smile and a word of kindness.

"Electricity at home brought about big changes in our daily lives. But there is one thing that will always remain and that is the quality of those memories from years ago and knowing that our men at the Farmers RECC are just as committed to keeping the quality of our lives intact today."

An early annual meeting, where the owner/users meet management to hear about plans for the future. (Farmers RECC archives)

FARMERS RURAL ELECTRIC
COOPERATIVE CORPORATION

Miles of Line:	3,481
Consumers billed:	23,537
Wholesale Power Supplier:	East Kentucky Power
Counties Served:	Adair, Barren, Edmonson, Grayson, Green, Hart, LaRue, and Metcalfe

ADMINISTRATION

Henry E. Gardner	1938 – 1943
Rodney Young	1943 – 1943
Charles M. Stewart	1943 – 1948
J.B. Galloway	1948 – 1988
Jackie B. Browning	1988 – 2007
H. Wayne Davis	2007
William T. Prather	2007 – present

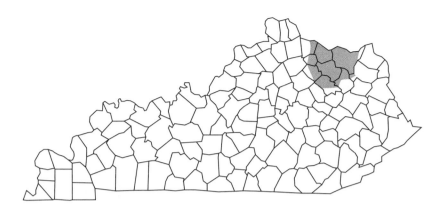

FLEMING-MASON
ENERGY COOPERATIVE

*I have seen the cooperative spirit alive. The men
and women who work in this program are salt of the
earth people and share a common bond.*

Christopher S. Perry
President and CEO
Fleming-Mason Energy

Old newspaper accounts, yellowed and tattered with age but filled
with names, dates, and memories, are cherished keepsakes, helping
today's fortunate citizen to understand and appreciate the earnest
beginnings of rural electric cooperatives.

The *Fleming Gazette* of February 29, 1940, is a good starting point.
In it W. Witten Horton, superintendent of Bath County schools, wrote:
"Where once was darkness, now is light. Those of us now enjoying
electric service for the first time are constrained to wonder how we ever
got along without rural electrification."

Emory G. Rogers, superintendent of Mason County Schools, added:
"A fairy godmother in the form of the Fleming-Mason Rural Electric

The first Fleming-Mason Board of Directors included C.J. Ross, Mrs. Lutie Whaley, D.D. Porter, Frank L. Hinton, R.O. Hord, H.L. Smoot, and Harry Lowe... then, as today, the Board is elected by the people and each representative is responsible to the people who are the owners and the users. (Fleming-Mason Energy archives)

Cooperative Corporation (REA) came to our assistance. Today all of the schools are provided with dependable electric service."

The 1974 Heritage Edition of the *Flemingsburg Times Democrat* carried the story of the beginning of Fleming-Mason RECC:

"It all began in 1936 when C.J. Ross read a story in the now defunct 'Pathfinder' magazine regarding rural electrification. He was reading the article by lamplight which was quite different from the lighting he had been used to in the city, before he returned to his family home. The next day when he went to his dentist, the late Dr. H.O. Dudley, who was then Mayor of Flemingsburg, and while his tooth was being filled, Mr. Ross questioned as to how he had managed a Federal Loan to build a new water system. Dr. Dudley suggested that he talk with the late J.M. McIntire.

"As soon as the dental work was finished, Mr. Ross went to Mr. McIntire's office to seek further information. As a result of the talk, Mr. McIntire agreed to go to Washington 'to see what he could do' with the REA. His fee would be $125.

"The fee was raised by contributions of $10 each from Mrs. C.R. Whaley, Reuben Tolle, Wilson Darnell, James H. 'Hop' Shanklin, Alex Reed, Frank Hinton, D.D. Porter, Robert Hord and Lloyd Tyree (the latter two from Mason County). Mr. Ross contributed the balance of $35."

"From that time on, after Mr. McIntire returned with the necessary information, the above group enlisted the aid of many Fleming and Mason Countians who worked hard signing up prospective members.

"On February 1, 1938, the first loan application was made to the Rural Electric Administration and it was signed by C.J. Ross, Mrs. C.R. Whaley, C.C. Rees, R.E. Tolle, and W.R. Guilfoile. It was for 97.9 miles of line with 320 members signed for service."

Names of thousands of others like these pioneers from the Big Sandy to the Mississippi represented determination, a common ground of hands and hearts reaching for the miracle of electricity. There'd be no turning back.

The raw energy emerging from Fleming-Mason was spreading through the Commonwealth. Other down-home cooperatives were just as eager in the universal plea springing from Genesis:

Let there be light:
and there was light.

Bible-directed but otherwise inexperienced individuals lacked funds and would need local and national leadership—the kind that had begun with President Franklin Delano Roosevelt, whose signature on May 20, 1936, established the Rural Electrification Administration—REA.

Investor-owned power companies would disagree, and some would cry "socialism," but the electric co-op revolution rolled on.

The first Fleming-Mason Board of Directors included C.J. Ross, President, Mrs. Lutie Whaley, Secretary-Treasurer, D.D. Porter, Frank L. Hinton, R.O. Hord, H.L. Smoot, and Harry Lowe.

A rich portion of the history of Fleming-Mason has been capsulated from the minutes of the organization:

July 22, 1937—"The first meeting of interested persons held at the Fleming County Court House with Dr. Earl Welsh, Agricultural Engineer from the University of Kentucky, as guest speaker. C.J. Ross was elected chairman of the temporary project committee; Mrs. C.R. Whaley, secretary; R.E. Tolle, Treasurer; and C.C. Rees and W.R. Guilfoile, members [Directors]."

February 1, 1938—"First loan application made to REA for $220,000 to build 97.9 miles of line to serve 320 members."

March 18, 1938—"Fleming-Mason RECC incorporated. Office located in the Wright building."

September 24—"First section of line completed between Fleming and Mason County by the Ray Chanaberry contractors. J.K. Smith employed by the firm as staking engineer."

1938-39—"Office moved to Browning and McKee Building on East Water St."

April 5, 1941—"First annual meeting. As of April, 1940 there were 1,483 members and as of Feb. 29, 1941—2,152. As of April 1940 there were 533 miles of line in operation and by Feb. 28, 1941 it had increased to 780 miles. Increased expansion was planned to include thirteen counties."

June 6—"Calvin Blair and William Jett approved as first full-time employees. Employees in 1941 beside those previously named included: Ruby Reger, bookkeeper ($110 mo.), Josephine Yancy, cashier ($85 mo.), Harriett Glascock, stenographer ($75 mo.), Ray Applegate, lineman ($140 mo.), Malcolm Sellers, lineman ($150 mo.), and J.K. Smith, manager ($235 mo.)"

Nov 7—"Voted to loan East Ky. RECC $300 to do preliminary planning."

1942—"Authorized payment to the State Association of Directors (probably beginning of Ky. RECC) on the basis of 5 cents per member, 2,234 members."

April 4—"First mention of KRECC as it is today. A merger of Association of REA Directors and State Association of REA superintendents and Managers. First mention of East Ky. RECC formed for the purpose of transmitting current to twelve member co-ops."

Nov. 19, 1947—"J.K. Smith, the first manager, resigned as of Jan. 1, 1948 to become manager of Ky. RECC in Louisville."

During 1949, the cooperative's territory was reduced in size with portions going to Grayson, Licking Valley, and Clark County cooperatives.

Sept. 21—"Birth of East Kentucky RECC. The Board voted a loan of $1,000 to join eighteen prospective member Co-ops in getting East Kentucky RECC started."

Although Fleming-Mason weathered the problems of World War II, it was not until the end of hostilities that major growth began.

The June 1948 issue of the Fleming-Mason RECC newsletter, *Rural-Lite,* featured "Mr. and Mrs. H.B. Williams who celebrated their 65th Wedding Anniversary on December 27, 1947.

In each cooperative beginning there has been limited space. Each square foot must be marked, carefully managed, and maintained. (Fleming-Mason Energy archives)

"Mrs. Williams, known as Aunt Mollie by her many friends, is eighty years old, and Mr. Williams has passed his eighty-seventh year.

"The Williams are through using the old kerosene lamp, for they became an REA member on April 10, 1948! Aunt Mollie says she just does not see how they did without electric service during their sixty-five years of married life. They intend to keep the old kerosene lamp just to show their great-great grandchildren how farm homes had to be lighted before REA days!"

Early 1940s engineering department—L-R, Front to back, C.D. Moore, Louise Payne, H.C. Graham, Charlie Jones, Robert Mason Ryan, Billy Gorman, Donald Lee Faulkner, Morgan Flora, Jim Mineer, and Jackie Jett. (Fleming-Mason Energy archives)

Meet Omer "Mousie" Crouch and his wife, Beulah Manley Crouch ("We just call him 'Mouse'."), in their well-lighted, neat-as-a-pin home on June 17, 2007, on the edge of Sharpsburg in Bath County, just south of Fleming County. The lights first came on when Mouse was nineteen years old, in 1953, the year Dark Star won the Kentucky Derby in two minutes and two seconds. Mouse was one of ten children living without lights on Cow Creek, and later across the hill on White Oak Creek just north of Owingsville, county seat of Bath.

"You'd find your way in with a raised lamp light. The Aladdin Lamp was for 'company coming,'" Mouse remembers. "Have to trim the felt wick—tedious lighting." Mouse's father, Curn, "wouldn't talk about electricity. Brother brought it in, never cost more than $2.50-$3.00 a month. Brother drove a school bus. Chopped out corn when it was a hundred in the shade.

"Worked from daylight 'til dark. Stripped tobacco in five to six grades." Mouse stripped out the reds or tips. His big brother lived farther up the holler. "Had an above-ground cellar with a spring running through it. Had a hole in it with water running right through and we'd go down and dip it. We kept milk and butter in it; had one cow. Water in the washpan, which we heated on the stove. Take bath in half gallon of water. Saturday night bath. Raised tobacco on as little as three to four acres. Sold tobacco for a quarter. Cold irons. Brother Frank died of diphtheria, died at home on the kitchen table. He was thirteen years old."

Beulah Crouch reaches in her memory book, back to Bald Eagle Creek, the little community of Bald Eagle (now gone), which was between Sharpsburg and Owingsville, five miles out of Sharpsburg—school, church, and store. Beulah's father ran the store. "Nothing there now," she seems heartsick to say. Beulah stripped tobacco and chopped out corn. There was no well. "We packed water from next door. Wash water was from creek. Up at 5 a.m. Black iron kettle, made hominy in it.

"'One of you girls go get a fresh bucket of water,' it was frequently said. So, we'd go to the well and draw it with a chain. Had a battery radio...Daddy feared running the battery down. We'd keep the radio under the covers, but he'd hear it and tell us to turn it down. 'Turn that thing off and I mean right now.'"

Beulah earned enough money stripping tobacco to buy a used '39 Chevy for $115. "Could buy five gallons of gas for one dollar. Kerosene was ten cents a gallon."

Mousie and Beulah are stalwarts in the infrastructure of rural electric cooperatives—the people for whom Franklin D. Roosevelt intended REA to serve. It's hard to imagine a worthier cause.

A conversation with Helen and Rube Blevins in their south Bath County home, which is near their country store and the U.S. Post Office in Preston, Kentucky. The old C & O Railroad tracks are gone. *The George Washington* passenger train doesn't go through anymore on its way to the nation's capital. On this day, it's the weekend of Court Day in Preston, where the highway is lined with people trading knives, guns, dogs, and old-fangled relics like dusty kerosene lamps.

"Take us back to the time when there were no lights in the house or on the farm."

"Well," says Helen, "I was about probably, maybe ten years old when we got electricity. Course we didn't own the house—we were living in a tenant house. And when he put the electric in we had a light bulb in each ceiling with a long string, and we thought we were rich, really. But they were so bright to what we had had, and you could do your homework and see, because we never did see well—two outlets in the whole house."

"Do you remember, Helen, some of the hardships before electricity came in?"

"Oh yes, like building the fires—the wood and stuff like that. Before we got electricity, the landlord had got electricity and we got a coal oil lantern. And we had a heavy coal oil refrigerator, and we thought we were rich then. My mother would build the fire. And we'd wait until she got a little bit started, and then we'd get up and be around the fire in the kitchen. It was an old house, and it wasn't easily heated."

The gathering of homemakers at Mays Lick, Kentucky, in 1940 puts a bright light on new and better ways of cooking, baking, and bringing a measure of ease to what had been primitive kitchens. (Fleming-Mason Energy archives)

Young homemakers display the benefits of electricity: modern, dependable, healthful, cool, clean, economical, easy, safe, carefree, and fast. So many attributes from the beginning that are taken for granted today. (Fleming-Mason Energy archives)

"Rube, what do you remember about not having electricity on the farm?"

"Well, I remember when we had all coal oil lamps. We didn't have no electric. We had no water. We had to cut wood every night for the fireplace. We lived in an old house and when the sun shined you could see through the cracks in some of the rooms. We had outside privies. We worked. We farmed on the halves, my daddy did. Everything went good. We were all happy. We were poor people."

"We were poor but we had love," Helen hauls back and throws in.

"Can you remember, Helen, that very moment when the lights first went on?"

"I can remember, because we waited for it. We just waited like it was going to be a show or something. And it was a show to us, because we could see everything so much plainer."

"What was it like, taking the Saturday night bath?"

"There wasn't much privacy, because all our rooms were really bedrooms. It was hard, but we did have water. We took a bath, we had a washtub, don't you know. We took a wash rag. One light in each room was all we had when we got electric, just one bulb is all we had. But we were all tickled to death to get that."

"Do you think that young people today could stand it? How it was back then?"

"They would have a hard time. We were all raised up in hard times."

Eighty-two-year-old Bernice S. Porter, almost fifty-four years an employee of Fleming-Mason, sums up loyalty: "My thoughts and prayers will always be with Fleming-Mason."

FLEMING-MASON ENERGY COOPERATIVE

Miles of Line:	3,456
Consumers billed:	23,482
Wholesale Power Supplier:	East Kentucky Power
Counties Served:	Bath, Bracken, Fleming, Lewis, Mason, Nicholas, Robertson, and Rowan

ADMINISTRATION

J.K. Smith	1938 – 1948
Robert L. Crager	1948 – 1949
C.J. Ross	1949 – 1973
Houston Delaney	1973 – 1989
Anthony Overbey	1989 – 2007
Chris Perry	2007 – present

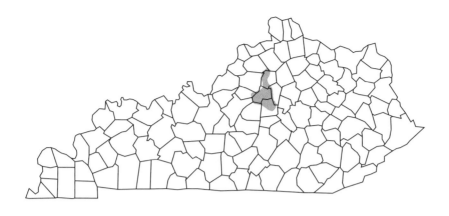

FOX CREEK
RURAL ELECTRIC COOPERATIVE
CORPORATION

The Board of Directors and I feel that consolidation
of Fox Creek RECC with Blue Grass RECC
is in the best interest of the membership.

Bob Kincer
General Manager
Fox Creek RECC

Fox Creek Rural Electric Cooperative was first organized in 1938.
The first headquarters was located in Stanley Trent's law office in
Lawrenceburg, Kentucky. Shortly thereafter, due to the need of more
space, the headquarters was moved to two rooms in Mrs. Gertrude
Ballard's building. The rent was $25 a month.

In "Yesteryear Glances, 25 Years Ago" in the December 12, 1963,
issue of *The Anderson News*, appears the following: "REA field agent
P.D. Carter, of Washington, D.C. was here surveying the rural electric
situation and said the project was classed by Washington as one of the
best in the United States. Carter said he was impressed with the work of
the resident engineer, F.L. Favor, the superintendent of the project,

George Sandidge, the spirit shown by the Board of Directors—especially did he praise the work of J.R. York, President of the cooperative, and Stanley Trent, the cooperative's attorney."

In the beginning at Fox Creek RECC, there were four full-time employees—Garey Wilmoth, Edwin Carter, Roy York, Gladys Sherwood, Manager George Sandidge, and five part-time employees.

The Anderson News story of 1963 also reported, "The history of the Fox Creek Rural Cooperative is a story of steady progress against seemingly insurmountable odds. It's the story of the determination of a group of people to obtain for themselves a service that had heretofore been denied them.

"The possibility of rural electrification was first discussed at a meeting of the Anderson County Farm Bureau, held in the County's Agent's office in Lawrenceburg on the evening of November 16, 1937. With the country in the throes of depression, rural folks were desperately searching for a means of increasing income and improving the rural standard of living. At this meeting, several committees were named to take the idea of rural electrification back to all the magisterial districts in the area to determine popular reaction.

"Matters moved at a slow pace until the night of March 22, 1938, when the first official meeting of the Fox Creek RECC was held. J.R. York was named acting chairman of the meeting and Elizabeth Toll Bailey, secretary. Other incorporators and members present for this meeting included I.B. Bush, W.R. McRay, W.O. Moffett, C.M.

A little girl who draws a name for a prize looks for a future brightened by cooperative togetherness. (KAEC archives)

Annual cooperative meetings attract user/owners by the hundreds, even thousands; it's a tradition, a gathering of community spirit. (KAEC archives)

Cornish, and P.H. Crutcher."

Groundbreaking for Fox Creek's permanent offices on U.S. 62 east of Lawrenceburg was held on September 17, 1962. Those participating included: Manager Roy York, President Floyd Watts, Vice President P.H. Crutcher, Architect Jack Clotfelter, Mrs. Aubrey Gritton, Marshall Warford, Secretary-Treasurer Frank Routt, Earl Dean Jr., Mayor John F. Lyons, County Judge W.W. Johnson, Mrs. Herbert Bowen, Earl Dean, Warren Wheaton, Representative Edwin Freeman, Mrs. Frank Routt, Mercer County Clerk Earl Young, Mrs. Garnet Carter, Mrs. Leroy Tracy, and Miss Ruth Slaughter.

"Architects for the building were Bayless, Clotfelter, and Johnson of Lexington. The contractor was E.H. Coulter and Son of Bloomfield.

"Fox Creek RECC has played its role in the social transformation of America's rural scene. This cooperative has provided hundreds of benefits, both social and economic, to the people in this area." (*The Anderson News*, December 12, 1963)

On September 23, 1997, Fox Creek RECC and Blue Grass RECC voted to consolidate effective January 1, 1998. Fox Creek RECC General Manager Bob Kincer explained: "The in-depth study shows that a consolidation of the two cooperatives will save approximately $12 million over the next nine years. These savings will be realized by the members of Fox Creek RECC in the form of rate reductions and the return of capital credits."

Dan Brewer, President & CEO of Blue Grass Energy, endorsed consolidation. "Both current organizations will be dissolved and a new company will be

It's never too late to be a winner, never too late to remember the "good old days" of kitchen drudgery when fires were stoked to bake a cake. (KAEC archives)

formed.... The new cooperative will be headquartered in Jessamine County and the Fox Creek office in Lawrenceburg will become a district office."

Although *The Anderson News* questioned the consolidation, the newspaper gave its editorial endorsement. Blue Grass RECC voted approval by a margin of 1,500 to 112. The vote at Fox Creek was 783 to 59 in favor of consolidation. The Kentucky Public Service Commission added its formal approval in an order issued December 12, 1997.

Fox Creek RECC's decision to consolidate with Blue Grass Energy in 1998 came as a result of a fundamental belief in unselfish cooperative effort, the spirit and the strength of rural electrification today. In 2002, Harrison RECC joined this consolidation with Blue Grass Energy.

FOX CREEK RURAL ELECTRIC COOPERATIVE CORPORATION ADMINISTRATION

George Sandidge	1938 – 1939
R.L. Wilson	1939 – 1940
Roy York	1940 – 1974
Charles Staples	1974 – 1985
Bob Kincer	1986 – 1998

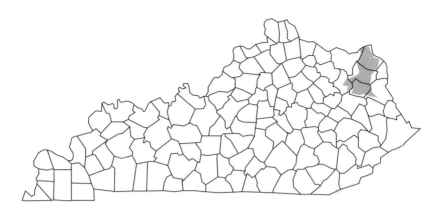

GRAYSON
RURAL ELECTRIC
COOPERATIVE CORPORATION

We believe our most significant achievement
has been to maintain a sincere sense
of community and concern for those we serve.

Carol Hall Fraley
President and CEO
Grayson RECC

Sitting and talking with Grayson RECC President and CEO Carol
Hall Fraley at her headquarters in the big white house at 109 Bagby Park
in the heart of Grayson—county seat of Carter—well, it's one of those
coming home feelings, warm, as they say, as butter cake. Even though
Thomas Wolfe said "You can't go home again," Carol Hall Fraley
believes in the realities of change. Grayson Rural Electric Cooperative
Corporation was incorporated in 1951 and moved into the Bagby
mansion in 1953.

"Mr. Bagby wanted to be a concert pianist—this was his summer
home. It was red Georgian brick, but one of the previous CEOs painted it
white—we had people to stand in the front yard and weep, but it's done."

The office of Grayson RECC is a place of pride in cooperative ownership. People working together make it happen. (Grayson RECC archives)

What else is done is electric co-op service to Carter, Elliott, Greenup, Lawrence, Lewis, and Rowan counties—2,425 miles of line up and down, back and forth, pole settings (sometimes by helicopter) through highlands of hope and plowshares of prayer that there'd always be light wherever needed. In Greenup County—W-Hollow home of the late author Jesse Stuart, *The Thread That Runs So True*—as of the summer of 2007, there were 15,699 rural electric customers.

But at mid-twentieth century along the banks of the headwaters of Little Sandy River—from one end to the other of unpresuming feeder streams like Doctor's Branch, Sheepskin Branch, and South Ruin Creek—the story of unlighted nights was the same as it was to the west, where Obion Creek and Bayou du Chien gave all they had to the Mississippi River. Whether in the south of the Cumberland River or in the north of the lower Licking, rural lines of electricity were virtually nonexistent.

The challenge has rested in the hands of community leaders like President and CEO Carol Hall Fraley and Board Chairman Roger L.

Trent. Problems are resolved in the house on the hill at Grayson Rural Electric Cooperative Corporation.

"By working together, we can make a difference and create a safer, more secure environment in the workplace, at home, and in school," says Carol.

"Working together" is a major mission statement, a guiding light at Grayson and the other twenty-five electric cooperatives throughout the Commonwealth.

Questions are frequently asked about higher rates—"Why are private investor-owned companies cheaper?"

"Because we serve six or seven houses on a mile of line," says Carol Fraley. "We serve where no one else wants to go. We serve those who want to live in the country and who don't want to live in an urban area. We're ninety-two percent residential...We see people on a weekly basis here who are coming in from other areas of the United States. I want to give back to our communities—it's a matter of doing more with less."

"Rate increases?"

"Rate increases are inevitable. There are a lot of small coal miners who are out of business—regulations did them in. Barge shipping is costly. Fuel, insurance, liability—things we have very little control over are costly."

And then there's the aching, emotion-charged issue of mountaintop removal. It's a reality moving closer to home with each blast of dynamite, each army of bulldozers.

"I guess we don't see as much of that as Big Sandy. I find it hard to fathom that they are going to move thousands of tons of dirt, and you're going to put it back to like it was.

Tom Martin of rural Carter County is proud of his Aladdin lamp, but he's more pleased that it's an heirloom he no longer uses to read schoolbooks and prepare his homework. (PLP photo)

But, if you can flatten it out, put homes, businesses, airports, etc., on that area, why not? If it's properly done, it can be done right. Before it was a goat path. We need our forests so that our children won't think that we're all blacktop, but we need places to build factories, highways, schools."

"Reclamation?"

"Reclamation is necessary—if that land is reclaimed so that it won't wash off into Little Sandy."

"What are some of Grayson RECC's community outreach involvements?"

"We have a place called Sara's Place, in Elliott County. We do interview skills with women who are trying to find employment, teach them how to interview, how to dress. These young men who work here, don't think anything about having a female boss. Those in their 30s just think that's the way it is. They've grown up with their mothers working, their wives work, so they don't think it so strange. Perhaps twenty-five or thirty years ago, it would have been different."

"What else?"

"Dealing with the poverty and the constant need to support and enhance education. We'll do anything we can—scholarships, speaking engagements, trying to keep them in school and to advance—it's a constant, constant battle.

"We have 16,000 meters, forty-five employees, not counting our college students or contractors, and 2,300 to 2,400 miles of line to substations."

Again, it's what J.K. Smith said about "starting with

The Buckeye School, built around 1915 in Carter County. One pole, one lineman, one connection to a one-room school in Kentucky waiting for rural electrification. (Grayson RECC archives)

people." So drive up the paved roads where once there were paths more accustomed to the plodding of horses and mules pulling wagons and sleds. Stop awhile. Visit awhile. Listen to the people.

Mrs. Leonard Adkins of State Route 7, Greenup, remembers the day the lights first went on along the headwaters of Little Sandy River. "It was very, very thrilling, really something else. My dad had a set of horses and he curled wire through the field. My husband had ten or eleven brothers and sisters. Seven brothers and sisters in my family. We had a wood stove. We'd go to the hills to get wood, coal for the sitting rooms.

"It was a miracle that came down through here. [Before electricity] we had a well with a hand pump. Put our milk and butter in a jar and put the jar in a bucket and hung it down the well. We had a big tub in the middle of the floor, which is how we washed. We took our time."

Come along for a visit with Jim Tom and Betty Martin. Jim Tom worked fifteen years for Grayson RECC. He started as "labor," then became groundman and lineman, did maintenance work for six counties. When lights began to come on about 1950, he was a senior in high school.

"I studied by Aladdin Lamp. The mantle gave more light," says Jim Tom as he reaches to a shelf to show a visitor. The Aladdin of Arabian Nights was fantasy; the Aladdin of the real world of Appalachian Nights was better than kerosene lamps, but light-years from the day of electric power. Or, so it's said. Jim Tom goes on.

"My grandmother and grandfather farmed to live. Had one milk cow or two. Raised tobacco and corn. Having electricity was like daylight from dark.

"We had an old well. Had a number 3 wash tub. Used a rope pulley at the well where we kept butter. A refrigerator was the first purchase after the lights went on. We had dirt and mud roads. Two miles as the crow flies from Grayson.

"Before electricity we had one lamp in every room—kitchen, dining room, living room, two bedrooms, two upstairs bedrooms. Had to be careful. Couldn't blow them out. You put in your hand, blew hard against your hand and that put out the flame. With the Aladdin you just turned it off."

Jim Tom Martin, seventy-four years old, was born in Grayson. His wife, Betty, is also seventy-four. He was raised by grandparents. "Had a good life. We were impoverished but happy to be so. We were very poor. It was the end of the Depression.

"Grandfather worked in summer, hunted in winter, fished and had a good time. Grandmother was very intelligent. She pickled corn and beans, had a potato hill, cabbage hill, and turnip hill. She packed straw around it. Didn't open a hill until she had to. The only meat was pork, which was salt cured and smoked along with chicken and wild game. We couldn't cure beef.

"We knew there was electricity [in town], but we didn't worry about it. In '48 and '49 the co-op was not in existence [in Carter County]. Our area was included in Fleming-Mason.

"We washed in a tub. We had a well twenty-five to twenty-eight feet deep. The wash tub had four wires. The well had a rope with a pulley four feet high to a cross member. We put milk and butter into the well— or anything that would ruin."

Jim Tom went to work at Grayson RECC in 1950—150 miles of power line. "We went to some hellish places. Contractors built lines, and we wired houses to make extra money. Refrigerator and washing machines were sold to the people being wired. Set meters. Good people. People so glad to get it. A small percent would bootleg electricity [jump meter boxes]. I went to work for a dollar an hour. Paid less than two dollars for kerosene. In 1950, farm work was four dollars an hour.

"When electricity came we were glad for the wire. It was pretty poor country. A pull chain light was all we could afford. Dairies sprang up. Stripping rooms had lights. Electric water heaters were later. With a well, it's either bad water or not much water."

"Taking baths—how was that?"

"Hard—especially if you were a long-legged boy. Used a number 3 tub, which was the biggest. Heat water on coal stove...get it real hot...pour into tub...add cold water...all done in kitchen. Nobody dared

Another annual meeting, another brighter day after the lights went on. (Grayson RECC archives)

come in. We all used the same tub, sometimes the same water. Summer time not bad…winter was different…wood and coal stoves in kitchen…fireplace in dining room, living room…Warm Morning coal heater. Bedroom upstairs, so we cut a hole in the ceiling for heat to come up. It would hit the ceiling and bounce back down if you were lucky."

Jim Tom's wife, Betty, talked about household "kettles of fish," sometimes Godawful realities. "You put sweet milk in the churn and set it next to the fireplace…let the milk sour and clabber. Mother would sing as she churned:

Churn butter churn,

Churn butter churn.

Peter's at the garden gate

Waiting for his butter cake.

"After the milk clabbered, you put the butter in a bowl and worked it down — slap it around. I was seven or eight years old," said Betty. "Walked one and a half miles to a little one-room school with nothing electric.

"My mom had a putt-putt washing machine run by gasoline. It was a happy day when the electricity came. No more outside toilet!

"We didn't have anything electric at the Everman Creek Missionary Baptist Church." Betty drew her strength from the strains of *Old Rugged Cross* and *Amazing Grace*.

Walter Sullivan, Sherman Gollihue, Dale Adkins, and Earl Hall. Four men and a spool of hope, convenience, and a better life on the farm. (Grayson RECC archives)

Betty has COPD (chronic obstructive pulmonary disease). She has an oxygen tank plugged in by a fifty-foot cord to the electric outlet in the wall, her "survival kit. If that goes off I'm in trouble," says Betty with a smile. It was never easy to rain on her parade. She was rooted in the place she loved.

"My mother was spotless. She had a four-burner Warm Morning stove. Had a side reservoir for water. Up over top was a warming closet to keep food warm. Bought kerosene for lamps. When you went to church you used a lantern or flashlight. No night light. I'm known for cream candy, stack apple cakes (eight-high, fried apple pies), and telling dirty jokes."

The smile broadens.

"I make blackberry jam...belong to Homemakers...240 pies gone in two hours. Sell 120 pounds of candy for four dollars a pound."

The Appalachian folksong *Churn Butter Churn* is a sweet memory. So is Betty Martin. She passed away February 4, 2008. Her final resting place is across the road up to the top of the hill in the family plot in Kemper Cemetery. Sweet Heaven, you might say.

Today's butter cake is reflected in even better light by Grayson

RECC. Carol Hall Fraley and her seven-man Board of Directors meet the fourth Friday of every month to fulfill their responsibilities, some required, some created.

"We have joined other Kentucky Touchstone Energy Cooperatives in offering innovative new programs to educate the community year-round," says Carol. "Grayson Rural Electric representatives visit schools and civic groups throughout the year, giving free presentations and demonstrations about electrical safety. Young children learn about basic safety around wall sockets and appliances, while older children and adults get information such as avoiding power-line contact. Demonstrations and lively discussions help drive the safety message home.

"We also have a Safety Trailer for outdoor demonstrations. Geared for older children and adults, this is an exciting demonstration about how electricity travels through wire and what happens when a kite hits a power line. Grayson Rural Electric provides Web-based tools to increase electrical safety awareness. Children love our Safety Station, where animated characters Buzz and Sparky teach them lessons in safety."

Churn butter churn!

Doris June Hall of South Shore remembers so well "the day the electricity was turned on in our home on Laurel in Greenup County.

"I was about twelve years old and we never had a Christmas tree. Mom and Dad never did the stringing of popcorn and the things you hear that others did to decorate their tree. That year our teacher at our school took down the tree early for Christmas break.

"She asked if I would like to have all the tinsel balls and rope from the tree. I was so excited I ran out and got me the prettiest little tree I could find from the hill. My brothers helped me set it up and I decorated it.

"But the tinsel didn't show up much in our house with oil lamps. Then the RECC of Grayson turned our electric on. Boy, did that tree shine! I had never seen such a beautiful tree with the foil glittering in those lights.

"I've never forgotten that tree, and now with all the electric lights we put on our tree, it can never compare with my memory of that little tree and our new electricity."

GRAYSON RURAL ELECTRIC COOPERATIVE CORPORATION

Miles of Line:	2,444
Consumers billed:	15,699
Wholesale Power Supplier:	East Kentucky Power
Counties Served:	Carter, Elliott, Greenup, Lawrence, Lewis, and Rowan

ADMINISTRATION

Hobart C. Adams	1950 – 1961
R.W. Thompson	1961 – 1962
Harold A. Haight	1962 – 1988
Mike Kays	1988 – 1991
Wayne Carmony	1991 – 1994
Carol H. Fraley	1994 – present

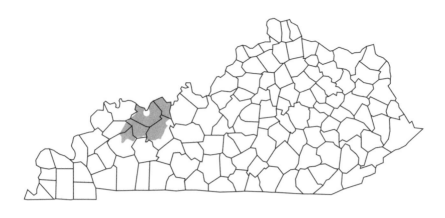

GREEN RIVER
ELECTRIC CORPORATION

The foresight, tenacity, risk-taking and imagination of its founders
coupled with the loyalty of its consumer-members
and skills and dedication of its employees—past and present—
have made the cooperative a success.

Dean Stanley
General Manager
November 1, 1984

Green River Electric Corporation was incorporated on June 11, 1937, and was consolidated in the formation of Kenergy on July 1, 1999, but the bright shining light that once bore the name Green River EC is still alive today.

Green River Electric, along with Henderson-Union and Meade County RECC, joined forces to form Big Rivers for the sole purpose of supplying wholesale power to the three systems. Big Rivers Electric was incorporated in 1961.

Like Fox Creek RECC and Harrison RECC, Green River Electric decided in favor of consolidation to become more efficient and effective organizations. Having said this, it's important not to forget beginnings. Green River Electric Corporation published in November 1984 *A History of Phenomenal Growth*:

"Green River Electric was conceived in March 1937 in the Daviess County Courthouse office of County Extension Agent Jack McClure. Birth followed in May 1938.

"McClure, who became one of the area's early leading pioneers in rural electrification, recalls that it was almost closing time when J. Warner Pyles of the REA office in Washington, D.C., appeared in his office that March day and broached the subject of a rural electric cooperative in Daviess County.

"On March 20, Pyles returned for a meeting with about fifty farmers, half of whom had not been on McClure's invitation list, but had learned of the proposed project and were interested. One of these was John Dawson, who figured prominently in the history of Green River Electric until his retirement at age seventy in 1965—first as a staking crew leader for the construction company and later as GREC's

> **July 18, 1947**
>
> **"Farm Family Day"**
>
> Oscar L. Camp, Director
> L. W. Crutcher, Director ------------------- Program Committee
>
> 9:00 A.M. to 11:00 P.M. .... Open House and Tour of Exhibits
>
> 10:30 A.M. Introduction to Cooking School ....Mrs. R. O. Brooks
>
> 10:30 A.M. to 11:30 A.M. Cooking School . Miss Marcella Liebeck
> Westinghouse Company
>
> 11:30 A.M. to 1:00 P.M. Boxed Lunches by Daviess County Home-
> makers Council
>
> 1:00 P.M. to 3:30 P.M. Demonstration of Farm Equipment
>
> 3:30 P.M. to 4:00 P.M. Kids Quiz—WOMI—Broadcast from Stage
> Main Auditorium
>
> 4:00 P.M. to 5:30 P.M. Musical Interlude
>
> 5:30 P.M. to 7:00 P.M. .... Snack Hour
>
> 7:00 P.M. Introduction to Cooking School .. Miss Katherine Shelby
>
> 7:00 P.M. to 8:30 P.M. Cooking School .... Miss Marcella Liebeck
> Westinghouse Company
>
> 8:30 P.M. to 10:00 P.M. Turtle Derby ----------------- JayCees
>
> 10:00 P.M. Announcements -------------- J. R. Miller, Manager
>
> 10:00 P.M. Awarding of Washing Machine by Green River R.E.C.C.
>
> SOFT DRINKS — SEE THE JAY-CEES BOOTH

consumer relations representative.

"County Agent McClure became the catalyst. His office became the headquarters and he directed the volunteers who went from farm to farm signing up members and securing rights-of-way for the power lines.

"The speed with which McClure directed the initial membership drive is attested to by the fact that when Pyles returned to Owensboro on March 30, a total of 1,056 signed survey/membership sheets were on McClure's desk.

"At that March 30 meeting there were a few farmers from Ohio and McLean counties present, who indicated that about three hundred of their neighbors also were interested. It was then that the decision was made to extend the project into McLean and Ohio counties.

"On May 21, 1937—two months after the historical courthouse meeting—Washington, D.C. [REA] announced the allocation of $100,000 for the construction of one hundred miles of line to serve about three hundred customers in Daviess County. It was added that the allocation eventually would be increased to a total of $292,000 for 280 miles of line to serve 1,122 customers in Daviess and McLean counties.

"With an attorney, documents were prepared and the cooperative was incorporated on June 21, 1937, by Howard E. Daniel, Thomas A. Cecil, Grover C. Wilson, Linnie W. Crutcher, and Oscar L. Camp, who formed the first Board of Directors.

"Membership recruitment was not as easy as the early numbers might have made it appear. There was a lot of skepticism, doubt and superstition. Many farmers also wanted to know how much it was going to cost before signing an agreement. Then, too, jealousy and dislike for neighbors entered into it. Some farmers would not agree to a line running across their farm that also would serve a neighbor they did not like.

"These were just a few factors that made going rough in the early days of Green River Electric.

"While the 'founding fathers' were all farmers and knew little about rural electric cooperatives, not to mention electrical engineering, they made up in dedication and hard work what they lacked in knowledge. This fact can be verified by reviewing the minutes of the numerous long and tedious regular and special meetings these men attended in the spawning years.

"During the 1960s, Green River President and General Manager J.R. Miller was a leading force in the more than $200 million industrial development of Hancock County. This development included the location of seven major plants within the county.

"The location of Harvey Aluminum [now Commonwealth Aluminum] in Lewisport was a lost cause until J.R. Miller said he could and would supply the electricity needed for the operation of the plant," said Hancock County native C. Waitman Taylor of Owensboro. "From this point, everything was positive. It was the turning point for the industrialization of Hancock County and J.R. Miller was primarily responsible.

"There was a chain of reaction once J.R. got Harvey Aluminum located," Taylor added. "J.R. was responsible for putting together the National Southwire Aluminum deal also.

"Throughout his years at GREC, Miller was active in the Kentucky Association of Electric Cooperatives and served on its Board of Directors for a number of years. He also was involved in the work of the National Rural Electric Cooperative Association.

"Miller also spearheaded the effort for the KAEC in obtaining a certified territory law through the Kentucky General Assembly in 1972.

"After repeated efforts failed to gain additional leeway in serving industrial loads and accessibility to purchase power, Green River Electric had no alternative but to continue plans for a joint venture with Meade County and Henderson-Union RECCs. The 'industrial revolution' had arrived in western Kentucky and industrialists and developers were inquiring of rural electric cooperatives in the area about adequate supplies of power."

Thus would be born Big Rivers.

Consolidation of Green River Electric and Henderson-Union Electric carried by a nine to one margin at Green River with a vote of 11,346 to 1,283. The vote in the Henderson-Union service area voted 4,478 to

3,182 for consolidation. (A first attempt at consolidation had failed.)

Green River Electric and Henderson-Union Electric became Kenergy Corp. on July 1, 1999.

It was estimated that the new consolidation would save $1.75 million to $2.5 million each year.

The cooperative's headquarters is located at 6402 Old Corydon Road in Henderson with service centers in Owensboro, Hanson, Hartford, Hawesville, and Marion.

down the line

PUBLISHED BY GREEN RIVER ELECTRIC
OWENSBORO, KENTUCKY 42301
APRIL-MAY 1981

MILLER RETIRES AFTER
35 YEARS AT GREC

J.R. Miller, like J.K. Smith, is synonymous with having a dream and working to make it come true. J.R. Miller was Green River Electric. (Kenergy archives).

GREEN RIVER ELECTRIC CORPORATION
ADMINISTRATION

Charles T. Smith	1937 – 1944
L.O. Denhardt	1944 – 1945
John Bishop	1945
L.A. Ehmsen	1945 – 1946
R.D. Osteen	1946
J.R. Miller	1946 – 1981
Dean Stanley	1981 – 1999

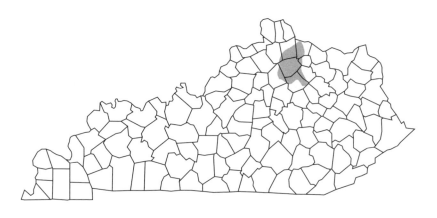

HARRISON
RURAL ELECTRIC COOPERATIVE
CORPORATION

In the mid-thirties, people who lived in Cincinnati
and Lexington could have electricity, but the rural people couldn't.
We wanted it, and we got it—through our own efforts.

James S. Patterson
President
Harrison County Rural Electric
Cooperative
About 1964

"Back in the mid-30s, the K.E.D. power company had a line two and a-half miles from Tricum [pronounced Trick-um, named for the belief that farmers were being tricked at market on agricultural prices]. People in and around that community went to the power company to ask for electrical services to their homes and farms. One of them was W.A. Penn of Renaker in northwestern Harrison County." (The Story of Harrison County RECC as told in *A Quarter-Century of Service* [1963])

"'They wanted $500 from each of us,' Mr. Penn recalled, 'and we would have to put in a lot of equipment or guarantee a big bill. We couldn't afford it.'

"Joe Marsh of Shady Nook had installed a $500 home electric plant that was expensive to run and gave him little more than minimum lighting.

"When, therefore, articles began to appear in the *Cynthiana Democrat* and *Log Cabin*, about a new government program called REA, Mr. Marsh and Mr. Penn read them carefully. And when County Agent H.H. Thompson called a mass meeting in Cynthiana for March 22 [1937] to hear an REA speaker, Mr. Marsh and Mr. Penn came early.

"The hall was packed. The man from Washington explained what was needed: a local organization, qualified under Kentucky laws to go into the power business. There had to be at least 150 people who would pay $5 memberships and guarantee to pay $3 a month for electricity, and who could all be served on fifty miles of line.

"Not everybody was sure it would work. But the meeting overwhelmingly voted to give it a try. They set up a temporary organizing committee, to visit their neighbors and sign them up. Joe Marsh was chairman of that first group, and later served the new Co-op as President for several years. Others in the group included Mr. Penn plus T.W. Maffett of Indian Creek as Secretary, J. Lewis Judy of Oddville, T.A. Collier of Robinson, Kirtley McDaniel of Leesburg, and William R. Jennings of Connersville.

"Like the present Board, these men served without salary. Until long afterward, they didn't even get their expenses.

"While the lawyers prepared incorporation papers, these men started out on a sign-up campaign. They drove miles, talked long and patiently. 'Not everybody was for it,' Mr. Marsh reports. 'Some were afraid of it.'

"'Yes,' Mr. Penn adds, 'they said it would draw lightning, the poles would fall down and kill a cow. One man refused to sign in what he called a wildcat scheme.'

"But most people did sign.

"It took a long time, but in February, 1938, the Harrison County RECC—at this stage it seemed likely to be entirely within the county, but people from other counties soon joined—received its business charter. An engineer—Ray Chanaberry of Louisville—spotted the sign up on a map, which went to Washington. After lots and lots of

Imagine the all-electric kitchen after eons of time! Was the wood-burning stove designed to punish? Was the "sad" iron destined to last 'til the end of time? Give Mom a break! (KAEC archives)

correspondence and many telephone calls, REA approved a loan of $201,500 to build 184 miles of line in Bourbon, Bracken, Harrison, Nicholas, Pendleton, and Scott Counties.

"They hired Guy Bridwell, who had been interested in electrical work in Sharpsburg, as manager.

"Soon poles and wires and transformers began to appear around the countryside. Those who had scoffed or refused to sign came in to clamor for service. Construction went ahead through the winter, and the delays seemed interminable. But the great day arrived on February 18, 1939. With a public ceremony in which Joe Marsh threw a switch, electric service finally came to the rural people of this area. By the middle of May, the entire system was in service. It has stayed in operation ever since.

"Did Mr. Penn and Mr. Marsh and their co-workers ever get paid for the hundreds of miles they drove? 'Not a dime,' they say. Are they sorry? 'No! It's the greatest thing that ever happened here, or to this State. We are glad we had a part in bringing it.'

Each passing year is marked with new electrical construction, new and better connections from generation to distribution to transformers to the flicks of many switches. (KAEC archives)

"So are a lot of other people."

People like Sue Simms Kelly of Harrison County, who has written the following for *Let There Be Light* in "loving memory" of her parents, French and Kathleen Simms.

"Electricity came to our part of the country [Harrison County] in the summer of '47. I was ten years of age at this time [eleven in the family].

"Before electricity, Momma had a metal icebox and she would purchase ice two to three times a week, and we would put old raggedy quilts and dishtowels around the block of ice to prevent it from melting. Our laundry was done in a large metal wash tub and washboard using homemade lye soap. Momma had a three-burner kerosene cook stove and it was my job to walk a half-mile to the store to purchase kerosene. I didn't mind doing this because there would be money left over to buy penny candy for me and the others. We also used the wash tub for bathing. However, I chose to use a granite wash pan for my bath.

"On laundry day, in warm weather, Daddy would dig a hole in the ground for a fire to heat the wash water. My sister and I would sometimes give Momma rest from the scrub board by washing the lighter soiled items and we would also hang the clothes on the line to dry. In the winter, our hands would be red and aching from the cold. Momma and we two girls would put our hands in a pan of warm water to help warm them up from the cold. My older sister Loretta and I would draw water from the well using a rope and bucket. When it was raining, we would take two kitchen chairs and rope and string a clothesline between the two chairs. For ironing, there were flat irons heated on the kerosene stove. One time, two of the burners on the stove burnt out leaving Momma with only one burner for cooking and heating water until Daddy could get the parts to fix the burners.

"Momma never complained. She made the best of what she had to do with.

"It was my job to drive the cow to the barn for milking each evening. I would sometimes start the milking before Daddy came in from the fields. I was a slow milker and Daddy would tell me he would finish the

An annual meeting's cake-baking contest centered around an electric oven is almost always a smile-making event. "Would you like a second slice?" seems to be the question posed by these winners. (KAEC archives)

milking. In the winter we milked by lantern light. Our butter was churned at home. It was so good on fresh baked bread. One winter, we had scarlet fever and were quarantined at home. The pond was frozen over and Daddy took an ax and chopped ice, which Momma took and put in a wash pan of water. She would dip the rag in the cold water and wash our face with the cold water to cool our fever.

"We did our homework by kerosene lamps. One day, we children saw men working in the field by our house putting up a tower. We asked Momma what they were doing, and she said they were stringing wires for electricity. We asked what that was and she said we would no longer have to use kerosene lamps to light our home.

"In those days before electricity, Momma and Daddy worked hard caring for their children. They instilled in us the rewards of honest work and a moral ethic to live by. I am thankful for being able to have lived in the days before electricity and to see the great change it has brought to our country. Those days are a sacred memory in my heart and shall remain so."

The founders of Harrison RECC, like the founders of Fox Creek RECC, in time, would have a difficult decision to make. The advantages of consolidation outweighed the temptation to stay in deadlock in relative isolation

It's called pole top "preventative maintenance," perhaps better understood by saying, "It's much easier to repair a line on a bright, mild day than during a midnight storm."
(KAEC archives)

The home office of rural electric cooperatives is an extension of down-home necessity, the place where cooperation begins. (KAEC archives)

from other co-ops. The central concern would be built on efficiency and effectiveness in service to consumer-owners, wherever they lived. Harrison RECC, with all its rich history, would find a new home within Blue Grass Energy.

March 1937	First meeting of interested farmers
March 1938	State charter granted
August 1938	First loan from REA
February 18, 1939	First line energized
1942	New headquarters constructed
1942-1945	War scarcities make expansion impossible
1946-1949	Postwar boom
1947	Guy Bridwell, first manager, dies; W.O. Penn replaces him
1947	2-way radio installed
1949	Area coverage attained
December 1954	First power from East Kentucky Power

1964	Second quarter-century of service starts
2002	Harrison Electric Cooperative consolidates with Blue Grass Energy

HARRISON RURAL ELECTRIC
COOPERATIVE CORPORATION
ADMINISTRATION

G.L. Bridwell	1938 – 1947
William Otis Penn	1947 – 1978
Ernest L. Skinner	1978 – 1989
Danny R. Haney	1989 – 1996
Jack Goodman	1996 – 1998
Ken Carpenter	1998 – 2002

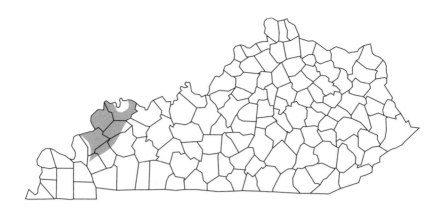

HENDERSON-UNION
RURAL ELECTRIC COOPERATIVE
CORPORATION

Line construction in the early days was time-consuming and difficult.
Holes were dug by hand. Poles were set by hand
with the aid of an A-Frame and winch.
Men, jeeps and mules pulled wire from reel jacks
to string wire across miles of line.
These hard-working men were often called
"Horsemen of the Lines."

John R. Hardin
Superintendent
Henderson-Union RECC

Henderson County RECC was incorporated on May 30, 1936, to become the first rural electrification association in Kentucky. This association paved the way for all other cooperatives that have since been organized in the Commonwealth.

The consolidation of the Henderson County and the Union County rural electrification projects became effective on December 1, 1939, with a combined mileage of 330, serving 653 members. The annual meeting

A historical marker commemorating the first electric consumer in Kentucky was unveiled on July 28, 1971, at Highway 41 Alt, six miles south of Henderson. The "juice" was turned on at Frank Street's Kentucky Cardinal Farms in October 1937. (Kenergy archives)

of the membership was held in the Farm Bureau building in Henderson on February 24, 1940. The following directors were elected:

Henderson County	**Union County**
R.R. Roberts	Richard Mills
W.C. Roberts	K.G. Davis
Rufus Eblen	H.A. Spencer

Officers selected by the Board of Directors:

President, R.R. Roberts
Vice-President, Richard Mills
Secretary-Treasurer, W.C. Roberts

From *The Connector*, Volume I, Number I, April 15, 1940, the
monthly news bulletin of the Henderson-Union Rural Electric
Cooperative Corporation:

"The office of the consolidated project is located in Henderson, at
215 North Elm Street, just below the Greyhound bus station. A pay
station is being maintained in Morganfield at the former location of the
Union County office. Mr. Ray Thomas, who is operating a furniture
store and repair shop there, will receive the payments of those who want
to pay in Morganfield between the 1st and the 15th of the month. Or
payment may be made by mail or at the office in Henderson.

"At the first annual meeting, J.L. Mudge, R.R. Roberts, Rufus Eblen,
Pruitt Priest, and Charles Duncan were elected as directors of the
association. F.J. (Boss) Pentecost was selected as the association's
attorney. He was instrumental in helping to draw up and prepare the
Rural
Cooperative
Corporation Act,
which was
adopted by
Special Session
of the State
Assembly in
1936 at which
time A.B.
(Happy)
Chandler was
Governor.

"The first
phase of the new
offices of the
Cooperative at
US 41 and 60,
just south of
Henderson,
opened the doors

*Together at the July 1964 annual co-op meeting were
(L-R) Robert Green, Senator Burdick (D) North Dakota,
special speaker for the meeting, Mort Henshaw, and John
R. Hardin, co-op Manager.* (KAEC archives)

for business July 1, 1947, and the second phase was completed in August, 1948. At this time the Cooperative had grown to 3,500 members in Henderson, Union, Webster, Crittenden, and Hopkins Counties.

"The year of 1948 was another year of expansion. Two-way radio equipment was installed, Mrs. Nina B. Gish was employed as Home Advisor, and in August, the Cooperative started billing by machines, instead of by hand.

"A $1,250,000 loan to serve 3,000 members and build 338 miles of line was approved by REA on January 1, 1949.

"F.J. Pentecost, the Cooperative's attorney since its inception, died August 6, 1951, at the age of 74, and on December 6, 1951, Rufus R. Roberts, the President of the Board of Directors since consolidation, also passed at the age of 65.

"The years ahead saw the Cooperative grow in miles of line, new and updated substations and the beginning of the Big Rivers G & T [Generation and Transmission] Cooperative. Frank T. Street was the first consumer of Henderson County Rural Electrification Association in October, 1937. He was the first member-consumer to receive rural electric cooperative power in Kentucky. Mr. Street owned and operated Kentucky Cardinal Farms on Highway 41 A, six miles south of Henderson. A Kentucky Historical Marker was unveiled at this location July 28, 1971.

"On December 5, 1938, Carl Ritz was employed as Henderson County RECC's first lineman. Electric wholesale power was purchased from the City of Henderson.

"Since January 1, 1940, our Cooperative has made remarkable progress in adding new members to the lines already constructed. More than 100 new members have been connected who were living along the present lines and were not receiving electric service. This period has increased our density from 2.2 member users per mile of line to 2.5, an actual increase of 101 new members in three and one-half months.

"During construction of the lines many transformers were hung and services built to people that had not wired their premises. They were put there with the expectation that they would be used, and the material represented quite an amount of money which was not being used.

Because of the increase in new members in the last three months, we have used practically all of this material, which was not being used before the next three months had passed."

C.F. Baker, inspector, of the State Electric Inspection Bureau, wrote in the first issue of *The Connector*: "At a meeting March 30, 1940, at the Kentucky Hotel, it was agreed by the inspectors to help the Cooperative out on additional wiring up to ten outlets or more. Instead of the regular inspection charge of $2.50 it is reduced to twenty-five cents an opening, so it will be to your benefit to be sure your wiring job is still O.K. and will stand inspection.

"One of your customers recently had a barn wired before inspected; in two days he burned 125 Kilowatts above his monthly minimum. This can happen to you if you let anybody wire in additional outlets. When extra wiring is done and is not inspected, your inspection slip becomes null and void. So please keep your wiring up to date and report all additional wiring for inspection at your Cooperative office."

J.R. Pike, a retired Henderson-Union employee, worked as

The duties of Frank Street, owner of Kentucky Cardinal Farms that produced tons of fine hybrid seed corn, were greatly relieved by electrification of his corn-grading equipment. (Kenergy archives)

a lineman from 1946 (after World War II) to 1985. He'll not be forgetting the day he was hit when 7,200 volts "in the top of the head... knocked me out for a day or two." With no hard hat, J.R. was working on a transformer after a storm, back in the days when pole holes were dug by hand and crosscut saws were the push and pull of aching muscles.

The story of Henderson-Union with all the peaks and valleys of growing up is a vital cornerstone upon which has been built a Kentucky-wide association

After WWII, electrical connections to the rural areas of Kentucky resumed and two young men who helped were (L-R) Jim Fred Mills (partially shown) and David Markham. (Kenergy archives)

of rural electric cooperatives. Each individual consumer-owner plays an essential role in all future development.

HENDERSON-UNION RURAL ELECTRIC COOPERATIVE CORPORATION ADMINISTRATION

R.E. Thomson	1938
John R. Hardin	1938 – 1949
John Hardin	1949 – 1975
Clark Jacobs	1975 – 1977
John West	1977 – 1999

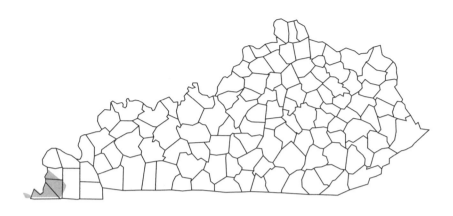

HICKMAN-FULTON COUNTIES
RURAL ELECTRIC COOPERATIVE
CORPORATION

Service is Number One—our only reason for existence.

Greg Grissom
Manager
Hickman-Fulton Counties RECC

In his office at 1702 Moscow Avenue on the hill overlooking Hickman, where the Mississippi River languishes but too often lurches toward its loop— New Madrid and Kentucky Bend—President & CEO Greg Grissom talks of hunting and electric-equipped duck blinds.

This may sound frivolous, but there's a positive economic benefit for an area that appreciates progress in a time of certain need.

"The popularity of hunting and fishing in Kentucky is indicated by the annual sales in recent years of more than one million hunting and fishing licenses," notes *The Atlas of Kentucky*. "In 1992, these sales generated nearly twelve million dollars in income. About fifteen percent of the licenses were sold to nonresidents."

Duck blinds in the Mississippi Valley flyway are only a fingerprint fragment of the special challenges facing Hickman-Fulton Counties

RECC. Other factors include a nationwide decline in the number of farms and troubling economic issues stretching from New York to San Francisco.

As Joe Creason would be the first to report with a respectful Jackson Purchase smile, it's important to understand that Hickman is the seat of Fulton County, and Clinton is the seat of Hickman County. The city of Fulton is in Fulton County, but never mind; part of Hickman County, better known as Wolf Island, lies west of the Mississippi River. Part of Fulton County is home to Reelfoot Lake, a little gift from the New Madrid Earthquake of 1811-1812.

Agriculture—mainly cotton, rice, corn, and soybeans—is at the weather's mercy, but the hold-on people who till the soil do not surrender. They keep working in the annual struggle to survive. Life isn't easy at the westernmost tip of Kentucky, which is true in most rural parts of the Commonwealth.

Barbara Lynn, whose family has farmed for years in Kentucky Bend, has the distinction of being the westernmost electric consumer in the Commonwealth. The 15,000 acres known as Kentucky Bend (you have to leave Kentucky and go into Tennessee, then back out again to get there) might be dismissed as the most unlikely of places for electrification. That assumption would be incorrect. What's good for Paducah should be just as good for the smallest toe of the Commonwealth.

"Alfred and Adrienne helped to get electric power in. He went around asking people, trying to get it up from Tennessee. He helped buy the poles," says Barbara, who with her family, the Lynns, carry on the Steppe family tradition of self-reliance.

Henry Cooley, eighty-five years young, a retired lineman for Hickman-Fulton RECC, remembers the early days of pole setting and wire stretching. "It was hard digging holes with the eight-foot split handle diggers and pulling wire by hand. Sometimes the ground was rocky and had to be chipped out. We didn't have much equipment. No bucket trucks, so we had to do all the climbing."

So what is the advantage of having three co-ops—Jackson Purchase

Those who served on the Board and led Hickman-Fulton RECC in 1979 were: (L-R seated) Joe Campbell; Harold Myers, Manager; G.H. Terry; and Paul Wilson; (standing) Charles Lattus and Larry Binford. (Hickman-Fulton RECC archives)

Energy, West Kentucky RECC, and Hickman-Fulton Counties RECC—serving an area as small as the Jackson Purchase—eight counties, about 2,000 square miles? Greg Grissom and the co-op he leads know people personally, more than just by a set of numbers—"A close presence when help is needed" is the way Greg sums it up.

Hickman-Fulton's relatively small size seems to set better with customers who harbor close family ties. The Kentucky electric co-ops are like that, all the more so when the miles of line are as few as 680 and the customers number fewer than 4,000, as they do in Hickman-Fulton, which today serves all or parts of Carlisle, Fulton, Graves, Hickman counties in Kentucky, and Obion and Lake counties in Tennessee. But, management remains open to the notion of progressive change provided it's best for the members.

Traditionally, one of Hickman's biggest employers is Hickman-Fulton RECC with fifteen employees and a payroll of $728,000.

The Hickman Courier, established in 1859, "oldest newspaper in Western Kentucky," headlined in the October 4, 1956, issue the "brand

121

new $75,000 headquarters of Hickman-Fulton Counties RECC." Co-op Manager Harold Everett invited all members to come in for an open house. "It's another sign of the growth and progress of Hickman-Fulton Counties RECC and the area we serve," said Everett.

"Hickman-Fulton Counties RECC was organized in 1938 and began delivering TVA generated power in June, 1939."

The newspaper quoted Everett: "The growth of the Cooperative since 1939 has been remarkable. Since then we've grown from a small co-op valued at only a little more than $100,000 to one valued at almost $900,000."

The July 9, 1987, issue of *The Hickman Courier* gave front-page coverage to the fiftieth anniversary of Hickman-Fulton RECC: "The first

Up in the buckets or on the ground are Jackie Curlin, Kevin England, and Chris Fuller in 2004 as they installed the three-phase transformer bank at Harper Hams, one of Hickman-Fulton's important clients. (Hickman-Fulton RECC archives)

step for obtaining rural electricity in this area took place in December of 1937. At this time a meeting was held with members of the newly formed Hickman-Fulton Counties RECC, representatives of the Tennessee Valley Authority and the Rural Electrification Administration, the engineering firm and the cooperative's attorney. A request was made for a contract with TVA for wholesale power to serve a section of members in this area to the Kentucky-Tennessee state line, north of Woodland Mills, Tennessee, and this was the first area to receive electric power.

"Also during December of 1937, by-laws were adopted along with legal papers necessary to secure a $100,000 appropriation. Work on the project was scheduled to begin in April and rural families were advised to plan wiring, decide on appliances they wanted and to make provisions for hooking up to the line as soon as the line was installed.

"At 5:30 p.m. on Friday, June 30, 1939, the first electric current was turned on in this area and additional lines were energized almost daily from that point on.

"Charter members of the Hickman-Fulton Counties RECC Board of Directors included: Roscoe Stone of Fulton County, President; J.B. McGehee of Fulton County, Secretary-Treasurer; and Harvey Pewitt. From Hickman County were E.C. Whayne, Vice President, and Grover Wyatt.

"Robert Hosmon was the first manager, Elizabeth Travis first bookkeeper, George Knight, first lineman, Edmond Wroe, first attorney, and Mary Beth Parker as the first Miss Hickman-Fulton County RECC."

Former managers include: G.H. Terry, George K. Knight, H.C. Schimmel, Harold Everett, John West, and Harold Myers.

"The History of Hickman-Fulton RECC"
from *The Rural Kentuckian*, 1987

"A dream which originated in the hearts of the rural residents of Fulton and Hickman Counties, took its first step in becoming a reality December, 1937. Lifetimes of endless physical drudgery and hardship were soon to be lifted of the burden and a miracle was about to take place—the day the lights came on. Electricity was going to change their lives forever.

"In January, 1938, Mr. Hosmon advised area newspapers that work was to begin in April. However, he cautioned, they would proceed carefully, as they could afford to make no mistakes. He advised rural families to plan wiring, decide on the appliances they wanted and make provisions for hooking up to the line as soon as it came.

"In February, an REA meeting was held at Cayce. There were 450 people in attendance. Mr. J. Warner Pyles of the Utilization Division out of Washington was the guest speaker. The main question among the people was, 'What will the rates be?' Mr. Pyles told them he thought the rates would be $1.00 for 25 kWh and $3.50 for 100 kWh.

"It was not until December of 1938, that John Gore from the Thomas H. Allen engineering firm was stationed at the REA office to stake off lines. There would be 70 miles of line in Fulton County and 35 miles of line in Hickman County. Construction was begun in January of 1939, by

Behind every board of directors, there's a team of technicians who know how to get the wires on the poles. Some of those on the team of Hickman-Fulton were Alan Wilson, David Craddock, Mickie Whitlock, Henry Cooley, Lloyd Baker, Johnny Davis, and Mr. Vowell. (Hickman-Fulton RECC archives)

R.H. Bouligny Inc. of Charlotte, North Carolina. The work began after fifty miles of right-of-way had been staked. Work was to be completed in early spring at a cost of approximately $722 per mile. Everyone wanting electrical service was urged to sign up at once.

"Materials required for the 105 miles of line began to arrive. Denkman Lumber delivered fifteen carloads of poles. The Aluminum Company of America provided 257 miles of wire and Westinghouse Electric furnished 212 transformers

These daring young men have no trapezes, or nets for that matter, as they attach lines to poles in the most southwestern co-op in Kentucky, an area adjacent to the Mississippi River. (Hickman-Fulton RECC archives)

with capacity ranging from 1 1/2 to 10 KVA. The construction of lines would provide 17,000 man hours of employment.

"As lines were being constructed, and the fulfillment of a dream came nearer, appliance shows were being sponsored by REA. They were held in various locations of Fulton and Hickman Counties. Refrigerators, washing machines, ranges, water heaters, irons, etc., were being purchased early in anticipation of the great moment when they would be put to use.

"Finally, at 5:30 p.m., on Friday, June 30, 1939, the first electric current was turned on. Mr. Luther Hutchison, from Fulton County, remembers that night.

"'I remember the day the electricity was turned on. That night, we got in our car and rode up and down the road to look at all the lights. Henry Maddox, down the road, had a big, two-story house and it was lit up like a hotel.'

The Pledge of Allegiance is a somber moment for young and old. To many it means loyalty and a belief in a nation that can cause there to be electrical power for everyone. (Hickman-Fulton RECC archives)

"The dream had come true and the hearts of many were filled with awe as they sat and gazed at those lights the first night they glowed."

In its 50th Anniversary report, 1987, Teresa Hayden recalled:

"I have worked for Hickman-Fulton for thirteen years, and yes, rural electrification gets in your blood when you are associated with it. However, after all this time, I don't think I ever fully realized the importance of this cooperative and the history associated with it.

"I guess a person my age and some even older than I, would take for granted just how comfortable electricity makes life and how convenient it is.

"I know that I owe a lot to the pioneers of rural electricity. They worked diligently and brought to reality a wonderful convenience that has spared me the exhausting physical labor that used to be spent just to provide the necessities of life. Clean clothes, food on the table and heat in the winter were not taken for granted before electricity came into being. Those pioneers and their vision have enabled me to read a book without excessive eyestrain, and hopefully, when I am old, I won't have the wrinkles under my eyes caused by years of squinting because of lack of light to sew or mend by. My children won't have chapped skin and coarse hands from carrying in buckets of water in the harsh winter weather and hot summer heat."

The Hickman-Fulton story has been repeated from Big Sandy to the Mississippi. Names and places may be interchangeable, but the theme is constant: before Roosevelt and REA, before rural people stood up to be counted, before the cooperative movement fastened its hold on the imagination of long-suffering people, there was darkness where there should have been light.

"Let there be light" was a cry that would not go unheard.

HICKMAN-FULTON COUNTIES
RURAL ELECTRIC COOPERATIVE CORPORATION

Miles of Line:	684
Consumers billed:	3,770
Wholesale Power Supplier:	TVA
Counties Served:	Carlisle, Fulton, Graves, and Hickman in Kentucky; Obion and Lake in Tennessee

ADMINISTRATION

Robert Hosmon	1938 – 1941
George K. Knight	1941 – 1943
H.C. Schimmel	1943 – 1948
Harold Everett	1948 – 1968
John West	1968 – 1978
Harold Myers	1978 – 1984
G.H. Terry	1984 – 1991
Glen T. Choate	1991 – 2000
Greg H. Grissom	2000 – present

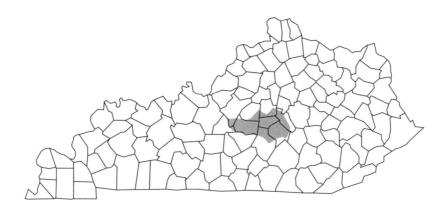

INTER-COUNTY ENERGY
COOPERATIVE CORPORATION

*We are not driven by profit, we are driven by our people; service
to our people is our driving force. We follow the Golden Rule—
we treat people the way we want to be treated.*

Jim Jacobus
President/CEO
Inter-County Energy
Cooperative Corporation

Drive up to the main office of Inter-County Energy Cooperative
Corporation—1009 Hustonville Road, in Danville—park your car, walk
through the co-op's front door, and there's an instant feeling of history
unfolding.

Pictures on the walls tell a story of patience, sacrifice, and
commitment to a bright idea—*light*—miracle to be shared, not
something exclusively for cities and towns.

"Anyone who walks through this door or calls on the telephone can
see me or speak to me," says Inter-County Energy President and CEO
Jim Jacobus.

"We're accessible. The phone rings directly to my desk and if I'm not there, they get direct voice mail and they get a call back from me. We don't play games and that sets us apart from other companies."

From the upper reaches of Rockcastle River to North Rolling Fork where it passes Penn's Store and Gravel Switch, Jim Jacobus knows there's just as great a need for electric power as there is, say, on the campus of Centre College.

"The future looks very bright. We are now at the stage of our lives, especially in Kentucky, where we have territorial laws. We have been patient, we've stood alongside and watched the other utilities get the large industries, but now the industries are in our territories—we are reaping the benefits of our patience.

"None of us wants to be a bad steward," says Jim Jacobus.

We sit together in a small conference room and are joined by Vice President for Customer Services Sheree Gilliam. We look back to newspaper clippings from mid- to late '30s. The source is the REA Edition of the *Danville Daily Messenger*, another example of the role of responsible rural journalism.

A.H. Karnes, county agent of Marion County, wrote:

"The first definite move toward getting Rural Electrification in Marion County was made in April, 1937, when sixteen leaders from over the county were called together and met with Mr. Earl G. Welch from the University of Kentucky Engineering Staff. At this meeting this group of leaders decided to ask Mr. J. Warner Pyles, a representative of the Rural Electrification office at Washington, D.C., to meet with a group of farm men and women on April 9. These leaders further agreed that if the project sounded feasible that they would make the necessary surveys in locating the customers and suggested and appointed individuals to survey each road in the county.

"At the meeting on April 9, attended by about two hundred farmers, it was definitely decided to continue with the project and George D. Harrison of Lebanon, Kentucky, was elected as chairman of the group to fill out the necessary forms. A survey was made and on April 25 a map showing the location of four hundred fifty customers on approximately one hundred and ten miles of line was forwarded to the Washington office with a request for the funds.

"About May 15 a notice was received that the Marion County project had been combined with Boyle and Garrard County and an allotment set aside from federal funds for construction of the line.

"At the organization meeting Mr. P.E. Hughes was chosen as director from Marion County and has served in that capacity since the formation of the cooperative.

"In discussing the history of the movement in Marion County, due credit should be given to Mr. George D. Harrison who served as the original chairman in the county group and assisted in preparing the original map. The Public Affairs committee of the Farm Bureau, composed of A.C. Glasscock, Sam C. May and J.W. Clark, along with Mr. P.E. Hughes, who was President of the Farm Bureau as well as Director of the REA, did all in their power to make the project work in the county and donated of their time and money.

"Perhaps it would not be fair not to mention the names of two women who stressed the importance of value of electricity in their daily contacts, namely, Mrs. J.W. Clark of Lebanon and Mrs. A.C. Glasscock of Penick."

When Inter-County families saw these three little words, "Removes Farm Drudgery," on the sides of these trucks, they probably appreciated them as much as another three little words, "You deserve better." (Inter-County Energy archives)

You deserve an electrified kitchen—a mixer, a stove, a refrigerator, oh, my! An exciting world opened up to the homemaker when the lights came on. Several "Halleluiahs" were heard along the farm roads of Kentucky. (Inter-County Energy archives)

Greetings from President Rogers
By W.H. Rogers
President

"We are very grateful to *The Messenger* for this opportunity of addressing each one of you personally. As the membership is increased in our cooperative, it is going to be increasingly difficult to maintain the close contact that is so necessary to the success of our project. We believe this Special Edition of *The Messenger* will do much to inform and inspire our membership.

"First, we wish to congratulate you on your loyalty, patience and spirit of cooperation. Whatever success our Cooperative has attained has been due to your loyalty and willingness to cooperate.

"Congratulations are also extended to the Kentucky Utilities who have cooperated in establishing a rate that would allow this and many other cooperatives to live. Kentucky Utilities not only has shown fine spirit but they have never contemplated building a 'spite line,' as has been done in many other states since the beginning of R.E.A."

Kearney Adams
Vice-President

"On June 15, 1937, a small group of farmers from Garrard, Boyle, and Marion Counties met in the office of the Farm Bureau of Boyle County and organized the Inter-County Rural Electric Cooperative Corporation for the purpose of supplying electricity to the farm homes of these counties. Beginning in these three counties with $150,000 with which to construct lines, we have reached out to ten counties and are spending more than $550,000 to put electric current into approximately two thousand homes.

"Staking of lines began about the first of December, 1937, with Richard Ethridge, representing Ray W. Chanaberry, Inc., as resident engineer. This was slow at first because the farmers could not understand the type of construction used, and many explanations must be made. Before each farm was crossed, the owner was taken out and shown that because of the long span construction, it was necessary to build the lines as straight as possible, with an angle when necessary instead of the short spans and curves he had been used to seeing. When told that all this was to his advantage, he readily saw the value of the type of construction used by REA two hundred and fifty miles of line for $199,000.

"The first lines constructed were in Boyle County. From Boyle the crew moved to Garrard County and later into Marion County.

"June 10, 1938, was a great day for the Inter-County Rural Electric Cooperative Corporation. One year after its organization electric current was to begin to flow for the first time into rural homes of Boyle County. The program began with a luncheon at 12:00 o'clock at the Gilcher

Hotel, attended by J.E. Van Hoose from R.E.A; the officers of the
Cooperative; J.V. Swaim, Project Superintendent; E.C. Newlin, attorney;
the office force; M.J. McDermott, President of Construction; and
Sylvester Pease, Superintendent of Construction; also Richard Ethridge,
from Roy W. Chanaberry, Inc.; John Brown, county agent of Boyle
County; B.W. Fortenbery, county agent of Garrard County; Ollie J.
Price, representing the Kentucky State Farm Bureau; and C.B. Hanna,
district manager of the Kentucky Utilities Company.

"After the luncheon the delegation proceeded to Perryville where
they were to witness the throwing of the switch that would energize the
sub-station and start the flow of current over the first line. As we all
looked on with quick beating hearts and bulging eyes, William Rogers,
President, closed the switch at 2:00 o'clock. As the sub-station began to
hum, we who had labored so long were almost as happy as they who
stood upon the banks of the Hudson more than one hundred years before
and saw the first steamboat begin to move. From here we proceeded to
Uncil Whayne's home where the first house was lighted. From this
beginning homes have been energized almost daily and many a heart
made happy because of the many uses that have been made of electricity.

"August 25th [1938] was a great day for the rural people of Garrard
County. On that day more than two hundred of them gathered at the
Lancaster High School to celebrate the coming of electricity to their
homes. They came carrying boxes, baskets, and bags filled with good
things to eat. These were received by a group of ladies in the cafeteria of
the school. The food was assorted and plates were filled for the crowd.

"E.C. Newlin outlined the progress of the R.E.A. B.W. Fortenbery
told of things to be done on the farm with electricity. J.V. Swaim said
that at that time Boyle County's one hundred and five members were
using ninety-six refrigerators, one hundred and forty-eight radios, one
hundred and thirty-two irons, forty-nine washing machines, eighteen hot
plates, and thirteen water pumps. Mr. Kelly, from the University, said
one great advantage of electricity on the farm was keeping boys and girls
satisfied there.

"At 2 P.M. the crowd had moved to the sub-station and with ears
tuned in and eyes set, watched the throwing of the switch and heard that

delightful hum of power that was to mean so much to them later. About fifteen minutes later, at the home of Mr. and Mrs. J.E. Johnston, on the Richmond Road, the lights were turned on, a radio began to play, and a dream had come true.

"Garrard County was then given the distinction of being the first county to have her quota of three customers per mile, wired at the time energy was received.

"After the Perryville and Lancaster substations had been energized, the people of Marion County still looked forward to their big day. Although they were a little late to receive the service,

*Before electricity came along, a crystal chandelier hung unlit for twenty-five years at B.P. M*c*Makin's, but when Inter-County's first switches were thrown, the fixture finally sparkled to life.* (Inter-County Energy archives)

they were as jubilant as the others had been. When the Lebanon station was energized and all lines turned on, current was flowing into homes in Garrard, Boyle, Mercer, Lincoln, Washington, and Marion Counties, over a network of two hundred and seventy miles of lines.

"When this is completed, our lines will reach farm homes in ten counties, namely, Garrard, Jessamine, Boyle, Mercer, Lincoln, Casey, Marion, Washington, Nelson, and LaRue. We are now working toward another fifty miles and where we will stop, no one knows."

In *Let There Be Light: A History of Inter-County Rural Electric Cooperative*, there's a description of life on the farm before electricity almost to defy belief. "Washdays were all-day affairs. You began by hauling enough water to fill a copper boiler and a couple of number 3 washtubs. You heated the boiler on the stove while in one of the

135

washtubs, you began removing a week's worth of dirt from your family's clothes with a washboard and a cake of lye soap. The scrubbing and the soap left your hands swollen and often bleeding. You put the scrubbed clothes in the boiling water and stirred them for fifteen minutes with a broomstick or heavy wooden paddle, then transferred them to a second tub for rinsing. White clothes required yet another tub for 'bluing' or bleaching. Finally, you wrung the heavy clothes out by hand and hung them on a fence, a tree or a clothesline to dry. With every load you repeated the process, including hauling fresh water, and as large families were the rule on a farm, a week's worth of laundry was never small.

"If you lived near a stream, you could reduce the distance you had to carry water and the amount of heat inside the house by heating the tubs over a fire along the bank. There was no way to make ironing easy, however. Irons in those days were made of real iron, and the only way to heat them was in front of a fireplace or on top of a stove. You kept at least two, so you could use one while the others were heating. You held each seven- or eight-pound iron with a thick potholder to avoid burning your hand. The irons collected soot and ashes from the fire and transferred them easily to clothes if you didn't wipe them constantly.

"Homes

An electrified brooder house brings good smiles and good results when biddies are happy and warm. In the early '50s, to run a chicken operation was pretty cheap. It's still a bargain.
(Inter-County Energy archives)

were hard to keep clean with smoke from wood stoves blackening walls and ceilings and only straw brooms to keep dirt and ashes off the floor. Saturday night baths in a No. 3 washtub meant hauling and heating more water. At night you lit your home with a coal oil lamp and probably could only afford to burn one at a time. In the morning you washed the lamp's

Mr. William H. Rogers (right), Board President of Inter-County RECC in the '50s, is overseeing an electrically-driven hybrid corn seed grader used on his farm. His smile tells it all. (Inter-County Energy archives)

glass chimney to remove the smoke from the night before. You also trimmed the wick so the lamp would burn properly. If you trimmed it too low, the lamp gave too little light; too high made it too bright to see by. Trimmed correctly, the wick produced about as much light as a 25-watt bulb. With no more illumination than this, Father read his Bible, Mother sewed and children did their homework around the kitchen table.

"Not having electricity produced not only a life that was physically hard but also conditions that caused a steady migration from the farm to the city. Poor lighting at home and at school made children's study difficult, and academic performance suffered. Poor sanitation from a lack of running water led to hookworm, dysentery and other disease, especially in the South.

"Just think of not having to fill up the old lamp, trim its wick and clean the chimney every evening so that the family may have light to read and study by at night. Just think of standing in the house and throwing a switch at the barn, throwing a full glare upon things there. Just think of having electric refrigeration right in the house, without the extra trouble of going to the old, dilapidated icehouse and digging the old soggy ice from the pond out. Think, too, of the electric fan that can be turned on in the sick room or when company comes."

INTER-COUNTY ENERGY
COOPERATIVE CORPORATION

Miles of Line:	3,598
Consumers billed:	25,159
Wholesale Power Supplier:	East Kentucky Power
Counties Served:	Boyle, Casey, Garrard, Larue, Lincoln, Madison, Marion, Mercer, Nelson, Rockcastle, Taylor, and Washington

ADMINISTRATION

Harry J. Achee	1937 – 1938
J.V. Swaim	1938 – 1941
R.D. Osteen	1941 – 1944
W.N. Jackson	1945 – 1966
C.W. Foley	1967 – 1991
Leo Hill	1991 – 1999
Jim Jacobus	1999 – present

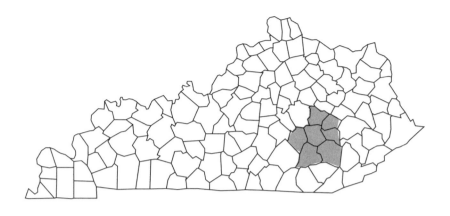

JACKSON ENERGY
COOPERATIVE

*The linemen are truly the modern-day version of cowboys
who brave the weather and ride the range of ridge tops and hollows
in bucket trucks to keep the lights on in a service area
that stretches through fifteen southeastern Kentucky counties.*

Don Schaefer
President and CEO
Jackson Energy Cooperative

The fifteen counties served in whole or in part by Jackson Energy Cooperative — Breathitt, Clay, Estill, Garrard, Jackson, Laurel, Lee, Leslie, Lincoln, Madison, Owsley, Powell, Pulaski, Rockcastle, and Wolfe — represent a total of 5,631 square miles.

It was in the summer of 1938 when Jackson County Rural Electric Cooperative Corporation was formed. Coleman Reynolds was the first President of the Board. The following year, the first electric line was energized near McKee.

In 1941, Board members voted to give each member of the co-op a table reading lamp, valued at $2.75, for each additional member they signed up for service. In 1945, Manager Moss Abshear reported he had held meetings in five of seven counties recruiting workers to sign up co-op members. Community meetings were held in conjunction with the County Extension Service. In 1949, at the Jackson County RECC Annual Meeting, co-op officials listed 6,653 members receiving electric service from 1,300 miles of electric line.

Karen Combs, Director of Public Relations for Jackson Energy, believes the "anecdotal and historical record of the transformation that occurred in rural Kentucky as a result of the REA program is a story more than worthy of the telling. It not only brought new technology to the region, but also transformed the way people lived, worked, and spent their leisure time. One of our retired employees tells an interesting story about an elderly gentleman who was skeptical of having electricity in his home. He thought it might somehow leak out of the light sockets and harm his family.

"Bringing electricity to the state also had residual and more personal benefits. Another story framed in our lobby is from the daughter of a Laurel County farmer.

It should never be forgotten that there was a time not so very long ago when the countryside was without lights. Nor should it be forgotten that the people own the co-ops that provide that light. (Jackson Energy archives)

Workers running electric lines in the area offered to pay the family to board with them while they completed their line work, and the extra income was a boost to the family at a time when money was hard to come by for many eastern Kentucky farm families. It wasn't just the benefit of bringing new technology to the region, but also the financial rewards that came from newly created jobs and the infrastructure construction.

"The advent of electricity also brought other home improvements, like indoor plumbing.

"The three most significant Jackson Energy achievements since 1938:

"1. Providing, maintaining, and transforming the physical infrastructure that brings electricity to our service area. The infrastructure now includes the latest technology, such as computerized GPS maps, satellite relays for metering equipment, and almost every co-op in the state has on-staff IT employees. ["IT" stands for "information technology," a term now used for all computerized activities of the co-ops.]

"2. Jackson Energy is a partner in the communities we serve. We provide economic development support to businesses not only through our physical equipment, but also by offering low-cost economic loans through RUS [Rural Utilities Service, formerly known as REA, Rural Electrification Administration]. We have awarded more than $60,000 in scholarships since 1992 and were the first electric cooperative in Kentucky to offer a voluntary community support program called Operation Round-Up. Operation Round-Up provides funds to nonprofit groups for community development programs.

"3. Hugh Spurlock, Jackson Energy's President & General Manager, was selected as the first President & General Manager of East Kentucky Power, the wholesale power supplier formed by area cooperatives to provide their own generation of electricity. Under Mr. Spurlock's leadership, East Kentucky Power built power

plants and was able to provide a reliable source of electricity for
its member co-ops.

"Jackson Energy will remain a member-focused electric cooperative
working to improve the quality of life in the communities it serves," says
Karen. "Fulfilling this mission leads to a variety of programs and
services, such as looking for new ways to help members conserve energy
and continuing programs that are already in place, such as Operation
Round-Up and economic development initiatives."

Mabel Arnold of Laurel County writes:
"My fondest memory of the Jackson County RECC came in 1952
when electricity first came to the community of Baldrock [in the
southern part of the Daniel Boone National Forest between the mouths
of the Rockcastle and Laurel rivers where they empty into Harriett
Arnow's long a-winding Cumberland River]. In 1952, times were hard
for me and my husband, Les Arnold. Cash was hard to come by and we
were raising five children. To supplement our income we would
sometimes take in boarders. Four such boarders were linemen from
Jackson County RECC. The men approached my husband and offered
him $1.50 a day for each man for a bed to sleep in and three meals a day.
"The men were extremely hard workers. They would get up every
morning at 5 a.m. I would feed them breakfast, then pack them a lunch
of either leftovers from the night before or sandwiches. At 6 a.m. the
men would leave for work and not return until dark. Evenings were
special with these men in our house. They would talk about the places
they'd been and how far they had gotten during the day. But mostly, they
would talk about their families, and how much they missed them.
"About halfway through their stay with us, the men encountered
problems getting their trucks to spots they needed to set poles. They
were working in awfully rough territory with the mountains going
straight up and straight down. The men offered my husband $1 a pole if
he would use his horse to skid the poles to where they needed to be. The
men would then pack their equipment by hand up and down the
mountains to dig the holes to set the poles.

First pole, first Board of Directors for Jackson Energy. Some of the members who helped raise that pole in 1939 were: Coleman Reynolds, D.G. Collier, R.H. Johnston, W.R. Feltner, George Sparks, L.H. Sparks, and J.R. Moberly. (Jackson Energy archives)

"These men only stayed with us three weeks, but they were like family by the time they left. I will forever be grateful to them for helping us through some hard times as well as for the electricity they left behind. I hear people talking about the good old days and how great they were, but for me, the good old days are now.... You can keep the good old days. I will keep my electricity and be forever grateful to the men who made it all happen—the linemen from the Jackson County RECC."

William Keith Combs III writes from Brushy Fork Road down about Berea: "I saw that you asked your readers to send in stories about electricity first coming to some areas in Kentucky.

"Well, I immediately started laughing about what my grandfather and my mother (his daughter) told us when we were children, along with many other good stories. My grandfather was Raymond Gay, my mother was Connie Gay, before she married my father, Bill Combs. Mama passed away last July 30th and Pa Gay died several years ago. Anyway, Sir, it seems that when my grandfather was a boy living in a holler way

"Sign me up!" And cashier Nina Anderson provided service with a smile for new electrical hookup meter requests. (Jackson Energy archives)

back in the hills of Jackson County, the 'lectric wars' [electric wires] finally got back to them in the holler—to all of the five or six homes within a mile of them.

"Everybody had finally got electricity but many people were really leery of it. For instance: (1) nobody would dare walk under the wires, and (2) many older people thought that if they could hear the voices on the radio, then surely the people on the radio could hear them.

"But it all came to a head that first winter when everyone caught the flu. Pa said it was very contagious and everyone was sick. (He always told this to be the truth.) Anyway, Sir, some of the elders figured out that it was the 'lectric' that was making everybody so sick. So everybody gathered the light bulbs, clocks, and radio, anything electric in a big pile, then they tied rags or handkerchiefs around their mouths so they wouldn't breathe nothing, then they proceeded to take axes and chop and bust everything to pieces, bagged it up in feed sacks, and buried it deeply underground.

"To top all of that, everyone started getting better. He said it was a long time before he could be convinced that the 'Lectric' and the 'Lectric Wars' didn't have nothing to do with the sickness that winter.

"My great-grandfather made the statement: 'I knowed that "lectric" weren't no damn good: hit'll be the death of ever head of cattle and anythang else that moves—clude'n us.'"

Jess Wilson lives on Possum Trot Road in northwestern Clay County. That's in the rugged region where Gum Branch, Ponder Branch, Pigeon Roost Branch, and Robinsons Creek trickle and unpredictably tumble. Jess remembers very well those days when there was no electricity in the area now lighted by Jackson Energy Cooperative.

"All the difference in the world," says eighty-eight-year-old Jess from his vantage point high above Gum Branch in Clay County down the road a ways from Egypt in Jackson County. Electric power "meant being a part of the world and not left behind," says Jess, one of the early workers and founders of rural electrification in the Commonwealth of Kentucky.

"Go back to the mid '30s— how difficult was getting drinking water?" a writer passing through would like to know.

"Got it from a well in the back yard and when that went dry, we had to tote from up on

Linemen go where ordinary mortals dare not. Rudy Fielding and his sidekick are some of the players who have had dangerous roles to fill on the stage of rural electrification. (Jackson Energy archives)

145

A horse, a rider, a pickup truck, and most of all, an over-arching idea to bring service to enlighten areas where once most outdoor work stopped when the sun went down. (Jackson Energy archives)

the hill. When the well didn't go dry, you had to carry it—a bucket in each hand. You used the ropes. We'd be used to the bacteria, but a cousin would come down from Ohio—they'd have the crud. Once they got used to it, well, you drew it from the well. Drew it up in a bucket."

"What about bathing?"

"Monday, Wednesday, and Friday—washed down as far as possible. Tuesday, Thursday—up as far as possible. Saturday you washed possible."

"Rural people, I'm told, have a reputation for watching every penny. Is that about right?"

"My daddy's first cousin got his nickname 'Stingy' because he was so frugal he would jump his paling fence rather than put wear on his gate hinges.

"When the electric came through, one woman came up wiping the tears with her apron saying, 'Thank God you've come,' or so the story goes."

And there was the woman reading her Bible when her children threw the switch for the first time—she thought the Lord had spoken, thought he had boomed, "Let there be light!" She let the light stay on until the filament gave out, and had to be reminded that from time to time even light bulbs must be replaced.

All this was before Interstate 75 was engineered north to south along the western edge of Jackson Energy Co-op's territory. With double four-lanes came development that multiplied the lights at Renfro Valley Entertainment. And today, the Wal-Mart Distribution Center in Laurel County is "a vital transportation center for the largest retailer in America," says Jackson Energy's CEO and President, Don Schaefer. "The music of Renfro Valley and the volume of Wal-Mart are treasured assets at the rural electric cooperative that serves them."

The area served by the interstate has grown and prospered, leading to Jackson Energy opening its first district office in London, Kentucky, in 1989.

Shortly after I-75 came through, the co-op had 21,042 members served by 3,594 miles of line. By 1996, that number had more than doubled to 43,500 members and 5,081 miles of line. Much of the growth came in Laurel County, which has two interstate exits. While the interstate brought growth along a major north/south corridor, the Hal Rogers Parkway and Russell Dyche Memorial Highway also opened up Laurel County to the east and west.

Jackson Energy Cooperative is hard at work continuing to bring light to all those homes and businesses along the main roads and far off the beaten paths.

JACKSON ENERGY COOPERATIVE

Miles of Line:	5,631
Consumers billed:	51,226
Wholesale Power Supplier:	East Kentucky Power
Counties Served:	Breathitt, Clay, Estill, Garrard, Jackson, Laurel, Lee, Leslie, Lincoln, Madison, Owsley, Powell, Pulaski, Rockcastle, and Wolfe

ADMINISTRATION

Ralph Skiff	1939
A.S. Atwater	1939 – 1940
Lester Reynolds	1940 – 1941
James C. Roby	1941 – 1942
James Jennings	1942
Hugh Spurlock	1943
Harry Tussey	1943 – 1944
Moss Abshear	1944 – 1946
Lester Reynolds	1946 – 1947
Hugh Spurlock	1947 – 1951
Luther Farmer	1951 – 1974
Lee Roy Cole	1974 – 1991
Doug Leary	1991 – 2000
Don Schaefer	2000 – present

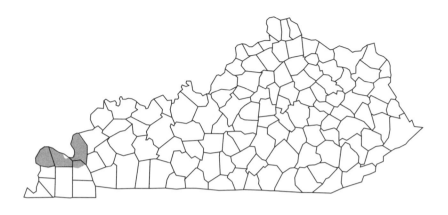

JACKSON PURCHASE ENERGY
CORPORATION

*A hallmark of JPEC is our competitive rates and superior service.
I encourage you to provide suggestions to improve your service.*

G. Kelly Nuckols
President/CEO
Jackson Purchase
Energy Corporation

The year of Jackson Purchase Energy's incorporation was 1937, the time of the Great Flood, when the Ohio River and its major Kentucky tributaries—Big Sandy, Licking, Kentucky, Salt, Green, Tradewater, Cumberland, Clarks, and Tennessee rivers—wreaked devastation and inundated Paducah. Estimated damage: a quarter of a billion dollars.

The floodwaters subsided in February, and in June Jackson Purchase Rural Electric was incorporated. The first directors were Horace E. Harting, Chairman; Roy J. Meahl, Vice President; and Harvey Lutrell, Carmel Harris, Ed F. Warren, Boone Hill, Claude E. Seaton, Roudell Wilson, and Walter O. Parr.

They saw the need.

Sometimes, the most fundamental of necessities go unattended until inspiration arises, and action replaces indifference. "People," said J.K. Smith. "It begins with people."

"One of the most prominent and significant elements of the history of rural electrification in Kentucky," says KAEC President Ron Sheets, "is how many prominent people were involved in the origin of the Commonwealth's electric cooperatives. One of thousands was the father of the Senior Judge of the U.S. Western District of the U.S. District Court, Edward H. Johnstone, W.C. Johnstone, who served as an Agricultural Extension Agent of McCracken County."

A feature article on W.C. Johnstone was published in the August 1, 1956, issue of the *Paducah Sun Democrat*.

"I helped initiate one of the greatest movements that ever came to McCracken County—the organization of the REA. One of the first meetings of that group and one of my last as a county agent was called at the home of C.D. Harris. There, with the help of Ben Kilgore, then secretary of the Kentucky Farm Bureau, and a few farm leaders, we started one of the first Rural Electrification Associations in the state."

While prominent individuals worked from the top, lesser known men and women were sweating in the kitchens of family farms.

"I have seen my mother stand over the wood stove

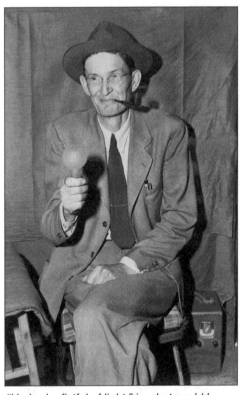

"He had a fistful of light," is what could be said of a very pleased Carmel D. Harris, early Manager of Jackson Purchase Energy. He also pushed the early fuse into place in April 1938 to energize the first thirty-five miles of line. (Jackson Purchase archives)

By 1946, the REA substation near Maxon had to be reconverted to handle a heavier load due to growing demands for power. Delivered and set up on January 19, 1947, each of these four transformers weighed 6,600 pounds. (Jackson Purchase archives)

while canning in the summertime with her face so red it looked like the blood would pop out," writes Dorothy Conder from Breckinridge County, upstream on the Ohio River from Jackson Purchase. "She had to can everything she could for food for the wintertime.

"When I was a child, we didn't have electricity in our home. My mother and daddy had a hard time making a living for their seven children.

"In 1955, I got old enough to get a job and have electricity put in our home. We had television—my dad liked to watch the game shows and the news, and it made life better for my mom. It made me happy to see them happy.

"Electricity did make a big difference in our home, and I am so glad I could do this for my parents—they did so much for me, and there was light in our home."

The counties eventually to be served by Jackson Purchase Energy— Ballard, Carlisle, Graves, Livingston, Marshall, and McCracken—were far removed and often forgotten by the urbanites in the golden triangle of Louisville, Lexington, and Covington-Newport as well as the cities of Henderson and Owensboro.

"We were in the dark before you came" were words often heard in farm homes, where women with hardened hands and aching backs lived out their shortened lives. The arrival of the pole setters, wiremen, and meter installers was as welcomed as if they'd arrived from another planet. They were greeted with smiles, and extra places were set at the kitchen table.

The new world of public, cooperative electric power would not have happened without people power. There were many heroes in the early Jackson Purchase days: John Fuller, Roy J. Meahl, Mrs. Allen Hines, W.C. Johnstone, Georgia Mae Nelson, C.D. Harris, Vilena Kelley, Alton Overstreet, J. Henry Kelley, J.E. Buchanan, Harvey Lutrell, Carmel Harris, Ed F. Warren, Boone Hill, Horace E. Harting, Claude E. Seaton, Roudell Wilson, Walter O. Parr, L.O. Solomon, Frederick E. Lackey. And there were many more who rolled up their sleeves and gave tirelessly of their time and talent. The mission was a constant "let there be light," and year after year there was light where there had been so much darkness. A new day had dawned.

The incandescent bulb was more than enough to put a smile on Carmel Harris's face. It brightened his day and made the tobacco in his pipe taste mighty sweet. A tattered scrapbook from the files of the Jackson Purchase REA includes stories from the *Paducah Sun* with a picture of the dapper Mr. Harris, long-time Manager of Jackson Purchase RECC, as he "pushed a fuse into place at the REA substation at Lone Oak...and the current flowed through the rural electrification lines. Thirty-five of the one hundred and twenty-eight miles were 'made hot' by Harris and J. Henry Kelly, REA lineman, before they were forced to quit work because of darkness."

As of March 8, 1940, Jackson Purchase REA had three hundred and twenty-two miles of line with one hundred and thirty-two more under construction or proposed and membership of about one thousand two hundred.

The Jackson Purchase archives reveal that in the summer of 1960 a

On the frosty morning of December 7, 1963, a tug boat and barge, along with co-op linemen serving as sailors too, laid a 3-1/2" submarine cable across the Tennessee River. It was laid in fifteen minutes due to months of planning to prepare for the day. (Jackson Purchase archives)

group organized under the name "JPRECC Members for TVA Power which favored applying for TVA power. Howard Reid of Symsonia was chairman for the Pro-TVA organization within the cooperative. On the committee were Leo Heidoran, William Hancock, Fred Smith, Guy Peck, Julian Hobbs, J.H. Davis, Jack Carroll, and Edward Reid.

"JP went to court and filed a lawsuit against TVA, asking the court to order TVA to supply electric service. The Co-Op's complaint was based on a Kentucky law which says that a power supplier can't deny service to a Co-Op which applies for it and agrees to pay going rates.

"December 30, 1962, after years of trying the City of Paducah got TVA power. JP remained the 'Island.'" But no RECC remains an island unto itself.

"Jackson Purchase Electric Cooperative began receiving power from Big Rivers on January 3, 1984."

June 2, 1997, Kelly Nuckols was hired as Manager of Jackson Purchase RECC.

An example of the pioneering role played by the Farm Bureau in the first days of rural electrification in Kentucky appeared on the front page of *The Hickman Courier*, August 27, 1936:

"A legislative program dealing with cooperation between the federal government and the state of Kentucky in regard to rural electrification, which is expected to be placed before the next special session of the General Assembly, was the chief topic of discussion Friday at a convention of Boards of Directors of the Kentucky Farm Bureau Federation in Paducah.

"The session, which was presided over by Ben Kilgore, executive secretary of the state farm bureau also decided unanimously to name the Rev. Parr, Paducah, Presbyterian minister, as organization director for Jackson Purchase farm bureaus.

Usually electric lines go up, not down, but in Jackson Purchase, lines must be laid and maintained wherever they must be. Here the first lines were being laid across the Cumberland River. (Jackson Purchase archives)

"Drawing executive officers and directors of the respective County bureaus, the convention was entertained at luncheon by the Homemakers Club of McCracken County at the bureau assembly house.

"Representing Fulton County were: Cecil Burnette, of the Fulton County Farm Bureau; Roscoe Stone, Vice-President; J.B. McGehee, Secretary-Treasurer; H.P. Roberts, J.R. Elliott and J.H. Lattus, Directors."

From the Jackson Purchase RECC archives comes a newspaper half-page of that era: "The job of supplying electrical power to the member owners of the Jackson Purchase Rural Electric Co-op is no different than the other...Electric Cooperatives in the state of Kentucky. But each Co-op like each individual has certain features that stand out as being different. The Cooperatives in the Eastern part of the State more often than not blow a hole to set a pole rather than dig a hole as they do in the Western end of the State. The rocky soil of the East calls for dynamite rather than a posthole digger.

"While each Cooperative is of a different entity they each have a common bond and feel deeply responsible for not only its own members but for the areas as a whole.

"The Jackson Purchase RECC has more than just a farming or industrial interest, as the membership is a varied one. In farming the membership ranges all the way from livestock—cattle and hogs to a bullfrog ranch—while another farmer raises sheep, another may have a poultry farm. The Co-op serves industry as varied as the farming enterprises. Mining fluorspar and limestone in Livingston County to clay in Graves County; while in Marshall County it's chemical industries that are a part of this Co-op family.

"In Ballard County near Monkey's Eyebrow, is a towering structure extending what seems to be miles into the air, a TV transmitting tower for WPSD Television. This TV tower carries news, weather and sports to more than three million homes in a four state area."

A Jackson Purchase publication reports that "In McCracken County the electricity for beaker lights and all the lighting system of Barkley Air Field is supplied by Jackson Purchase RECC.

"Eleven thousand members and 1,700 miles of electric lines all go to make up Jackson Purchase RECC, a Cooperative designed to do what an individual couldn't do alone. Designed to provide a service and serve a part of the total force to create greater development of not only our natural resources, but also our human resources—the true wealth of any County." (Jackson Purchase publications)

In his Annual Report for 2003, the President's message to the member-owners reads:

"The past year was a very good year for Jackson Purchase Energy. The company posted margins of nearly $1.6 million compared to $1.2

million in 2002. This further demonstrates JPEC's employees' commitment to do more with less. Improved efficiency and fiscal responsibility helped overcome lower sales in 2003 and bank a greater margin than in 2002.

"As the company continues to operate in the black and post margins, the question on many members' minds is, 'What happens to those margins?' The margins JPEC makes each year are used for updates and replacement of the existing electrical systems as well as new services and equipment. During 2003, JPEC invested $4.4 million in the extension and replacement of the system. In addition, $1.4 million was used to repay long-term debt obligations.

"The electric utility industry is complex. The electrical blackout and the widespread disruptions suffered by many consumers around Lake Erie and the northwest in August 2003 served as strong reminders of our dependence on electricity and how complex the business of providing electricity can be. JPEC remains committed to working with our power supplier and other cooperatives in utilizing hard work, technology, and cooperation to deliver reliable service at an affordable rate.

"JPEC's Trade-A-Tree Program allows for the replacement of trees coming in contact with primary power lines (those running next to the street, not the individual line to your house) with low-growing varieties such as dogwoods or redbuds. Please consider participating in the program and allowing JPEC to remove the offending tree. You will not only get a new tree in return, but you will be improving reliability for you and your fellow JPEC members."

President and CEO G. Kelly Nuckols, in his Annual Report for 2006, noted that "The end of 2006 capped a nine-year string of holding or reducing electric rates for our members. In 2000, JPEC lowered rates nearly three percent as a result of lower power costs. And, we reduced rates again in 2002. There are not many items we can say were lower in cost in 2006 than in 1997. Think back to 1997 and remember what a gallon of gasoline cost, the price of natural gas or propane, the price of a trip to the beauty shop or barber shop, or the price of a box of breakfast cereal. We are very proud of our efforts to provide our member-owners the best possible service at a very low and fair price. Today, JPEC's rates

When the lights go out, a substation might be down. Whom do you call? A generating and transmission facility that can provide continuing power through a mobile substation while construction, repair, or maintenance work is being completed. (Jackson Purchase archives)

are some of the lowest in our region, in the state of Kentucky, and in the entire nation.

"Technology can and has played a large role in reducing costs and we expect that trend to continue. Some examples include the future use of automated meter reading systems that, in turn, can help reduce the more than 50,000 gallons of gasoline we use each year. The use of computer-aided mapping is reducing the travel lengths to and from job sites.

"Not only is JPEC constantly searching for ways to reduce costs, but we also are searching for ways to help you reduce yours. For several years, JPEC has distributed high-efficiency compact fluorescent bulbs at our Annual Meeting and other community events. We offer a website full of energy-saving tips, and our customer service representatives are kept abreast of the latest energy-saving ideas in monthly training sessions."

JACKSON PURCHASE
ENERGY CORPORATION

Miles of Line:	3,252
Consumers billed:	28,710
Wholesale Power Supplier:	Big Rivers Electric Corporation
Counties Served:	Ballard, Carlisle, Graves, Livingston, Marshall, and McCracken

ADMINISTRATION

L.A. Solomon	1937
Carmel Harris	1937 – 1961
Hobart C. Adams	1961 – 1967
Howard V. Reid	1968 – 1969
James E. Campbell	1969 – 1982
John F. Ferguson	1983 – 1984
David Stiles Jr.	1985 – 1996
Donnie Lanier	1996 – 1997
G. Kelly Nuckols	1997 – present

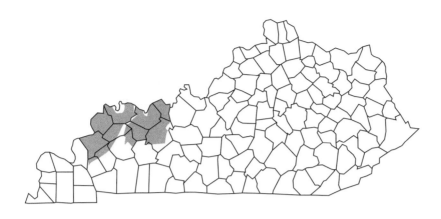

KENERGY CORP.

The co-op world is the best of all possible worlds.

Sanford ("Sandy") Novick
President and CEO
Kenergy Corp.

Kenergy, the newest cooperative name in the story of rural electrification in Kentucky, means to continue a tradition of joining together to build on individual deep-rooted backbone, an idea born in the years following the Great Depression.

Kenergy was established July 1, 1999, with consolidation of Green River EC and Henderson-Union RECC. The uniting was not without rivaling turf clashes—Henderson and Owensboro historically being two cities in close competition for markets, political influence, schools, and socioeconomic status. Consolidation demanded sacrifices of pride, bringing with it a larger concern over and above whose mutton was barbecued the best.

Dr. H.M. "Bo" Smith was the first Chairman of the Board of Kenergy after the consolidation and served from July 1999 to July 2002. "I supported the action due to the similarities of the two adjacent co-ops. The foresight demonstrated by those who advocated the consolidation has been vindicated time and time again. By combining, over two million dollars a year has been saved. This adds up to over eighteen million dollars since the consolidation."

Dr. Smith is presently a Kenergy Director. His mother and father, Lenn and Charlotte, celebrated their 71st wedding anniversary in 2008. Dr. Smith was born on October 19, 1939, "in the front bedroom of my then grandparents' home. The home was electrified in September, shortly before my birth. My mother remarked it was a 'Godsend.' She was referring to the coming of electricity, not my birth."

There was a time not so long ago when there were no electric lights on most Kentucky farms lying beyond the urban areas of Henderson and Owensboro, no lighted way up narrow hollers connected by bumpy roads, no running water or indoor toilets, no air conditioning when summer temperatures soared, no electric refrigeration in any season of the year. There were only wood burning stoves, scrub boards, and "sad" irons.

Dr. Smith's mother recalls "the drudgery prior to the installation of electricity, because all sheets and curtains were washed using a gas washer and gas ironer."

It was a time when rural people prayed, "Oh, Lord, let there be light," and it

A man named "Bo," Dr. H.M. Smith, Chairman of the first Board of Directors, affixed his signature bringing into existence Kenergy, the merger of Henderson-Union and Green River, while Attorney Frank "Nib" King and Dean Stanley, former CEO of Green River and first CEO of Kenergy, look on. (Kenergy archives)

was meant in a most literal sense. It is difficult to imagine that as late as only seven decades ago there were hundreds of thousands of Kentucky farm families without any electric power—no refrigerators, no washing machines, no running water, no air conditioners, no television, no computers, only coal oil lamps to read by. Some fortunate families had Aladdin lamps, which provided relatively brighter light.

The miracle of energy moving across the Commonwealth, lighting the landscape, alerting incandescent security lamps, is a reminder of how it was for Grandfather and Grandmother.

How could such a dark inequity be explained at a time when urban areas were so well-served by private, investor-owned power companies? As long as profit making led the lighted way, it was not feasible to establish costly electric service to a comparatively few rural customers living in distant, difficult places.

Yet, profit should not be demonized, nor should cooperative alternatives, when they are legally pursued. The Public Service Commission is the gatekeeper.

Presidents Calvin Coolidge and Herbert Hoover had opposed legislation that would have brought electric power to rural America. It was Franklin D. Roosevelt who set in motion the Rural Electrification Administration, so that finally there would be light where for so long there had been darkness.

From Big Sandy to Hickman-Fulton, Kentuckians such as those gathered in Henderson County in May of 1937 were about to experience a dream coming true.

Timeline of the formation of Kenergy Corp.:

June 4, 1936—Henderson County RECC was incorporated.

June 11, 1937—Green River Electric Cooperative was incorporated.

June 16, 1937—Union County RECC was incorporated.

October 1937—Frank T. Street became the first member-consumer to receive rural electric cooperative power in Kentucky. Energy was provided by Henderson County RECC, the first rural electric system in the state to be energized.

Kenergy's first Board meeting on July 1, 1999, and a show of hands for unified action. (Kenergy archives)

December 1, 1939—Henderson County RECC and Union County RECC consolidated with a combined total membership of 653, to become Henderson-Union Rural Electric Cooperative.

April 1946—J.R. Miller joined Green River Electric. He remained with the cooperative to see it become one of the most advanced rural electric cooperatives in the nation and was instrumental in the formation of Big Rivers Electric Corporation, formed primarily as a means to attract industry to western Kentucky. Industry meant jobs. Jobs meant payrolls. Payrolls meant economic development. Economic development meant release from hardship.

Green River Electric, along with Henderson-Union and Meade County RECC, joined forces to form Big Rivers for the sole purpose of supplying wholesale power to the three systems.

1960—Big Rivers was incorporated.

1972—Under the leadership of J.R. Miller, legislation was passed for electric utility territorial protection. Under this legislation, the Public

Service Commission has the right to establish certified service territories for all retail electric suppliers.

1997 — A merger proposal between Green River Electric and Henderson-Union Rural Electric was rejected by a narrow margin by the Henderson-Union members.

July 1, 1999 — Green River Electric and Henderson-Union consolidated to become Kenergy. As a result of this consolidation, a four percent rate reduction became effective for residential and commercial customers effective July 1, 1999.

2006 — Kenergy received the Kentucky Governor's Safety and Health Award for achieving more than 500,000 hours of work without a lost-time accident.

2007 — Kenergy was the recipient of the Industry of the Year Award by the Henderson County Chamber of Commerce.

2008 — Today, Kenergy is the third largest cooperative in Kentucky in terms of customer accounts and ranks No.1 in energy sales among electric distribution cooperatives in the nation, due primarily to large industrial usage by local aluminum smelters.

Kenergy serves in excess of 54,000 households, businesses, and industries along more than 6,500 miles of line in all or portions of fourteen western Kentucky counties — Breckinridge, Caldwell, Crittenden, Daviess, Hancock, Henderson, Hopkins, Livingston, Lyon, McLean, Muhlenberg, Ohio, Union, and Webster counties.

Kenergy's mission is "to safely provide low-cost, reliable electricity and related services not readily available elsewhere. Kenergy looks to maintain its position in providing its members one of the lowest electric rates in the nation. Kenergy's economic development professionals will continue to be actively involved with local economic development agencies to attract and retain businesses in western Kentucky. As it has in the past, Kenergy will continue to look for ways to enhance the lives of its friends and neighbors in the communities in which its members live and work.

"Kenergy will continue to educate its members on the importance of energy efficiency and demand-side management. It will promote the use of Compact Fluorescent bulbs, which use sixty-six percent less energy and last up to ten times longer than incandescent bulbs." (Kenergy archives)

In 2007, Sanford ("Sandy") Novick became Kenergy's new President and CEO. A former Tennessean with private and public power experience, he says he accepted the top job of management at Kenergy because "the co-op world is the best of all possible worlds."

Sandy Novick follows in the tradition established by the legendary J.R. Miller, who had come from the nation's capital and the old REA (Rural Electrification Administration) to provide professional leadership wherever it was needed.

A man with a vision, J.R. had a passion for economic development. For him, the connection with rural electrification was as natural as the flow of water from the springs to the branches to the creeks to the rivers to the sea.

Bettye R. Arnold lives in Cromwell in Ohio County near a loop of Green River. "I remember the well-lit days and the not so well-lit ones. I was the youngest of four girls and it was my job to chop the wood for Mom's wash kettle outside. Sometimes one sister would help. I don't know what the others were doing—maybe cooking dinner. But when it came time to wash the quilts, I was the one to get in the tub and stomp them clean.

"'If Abe Lincoln can study by a fireplace light, I can study by a kerosene lamp' was my thought. But the biggest change for me was the Aladdin lamp. My, what a difference it made.

"Back to the washing: the cows grazed where the hand-dug well was and when Mom got the washing done, she went to the house and I had to keep the cows away from the clothes on the line to keep them from chewing on them until they got dry."

When Randolph "Randy" Powell is not making decisions as a member of the Kenergy Board of Directors you might find him out in

There was a ceremonial ribbon cutting on July 1, 1999, to celebrate the grand opening of Kenergy's new office, which today serves more than 48,000 clients along more than 6,500 miles of line in a fourteen-county western Kentucky area. (Kenergy archives)

the calving barn on his farm on Whitelick Road, south of Corydon in Henderson County near the stream called Beaverdam. Randy has served his cooperative as a Board member for three decades. He remembers the times when Henderson and Union counties were stalwart competitors, strong enough that Union County broke away from Henderson County in 1811. In Kentucky, counties nurture individuality.

Almost two centuries later, Randy takes his cell phone with him to the barn as well as the boardroom of Kenergy Corp.

"I was one of seven children, four boys, three girls."

"How did you manage taking baths in so little space?"

"Lined up over in the corner with a wash pan, one after the other—boys one way, girls the other...had water in a big black kettle."

"Outhouse?"

"Two-holer—catalogues and corncobs."

"How did your daddy take to electricity? Did he have his doubts?"

"Had his doubts about the bathroom. He'd go outside. Couldn't bring himself to go inside. But he finally did."

"What was life like for your mother in the kitchen before electricity?"

"She was happy. Had a wood stove. Burned wood and coal. Coal-started fire would run all day. But, when Mother got her Maytag washer, she was on top of the world."

Eighty-year-old Norma Taylor Lashbrook writes from Philpot, Kentucky: "I would like to share the story of 'When the lights went on' in Habit, Kentucky.

"Habit is a village, ten miles east of Owensboro. My father and mother owned and operated the General Merchandise Store there from January 1929 to June 1950. The village was made up of the store, the Bethabara Baptist Church, (established in 1825), the blacksmith shop, Habit School and the telephone exchange— maybe fifteen houses with family farms surrounding.

"In the fall of 1937 our 'phase' of electrical power, from the Green River REA, was to be activated at 8:00 p.m. on a given night.

"We all gathered in the little square in front of the store and *waited!*

"Suddenly, at 8:00 p.m. there were lights in *every* home and building! It was so exciting—for heretofore we carried lamps from room to room, used iceboxes, heated

"What this church needs, preacher, is some more light!" and that's what happened in the latter 1930s when the old Bethabara Baptist Church in Habit, Kentucky, shone a new light on old text. Established in 1825, the church had spent more than 100 years searching for the light. (Photo by Debbie Hayden)

with coal or wood in grates or heaters. Most homes had been wired for electricity with only one drop light from the ceiling. Now we could use our entire house, both night and day.

From one-room, rural schools with outhouses, hand-pumped water wells, and potbelly stoves to schools with ceiling lights, water coolers, furnaces, and home economics classes with electrical appliances. Heaven on earth ... for teachers anyway. (KAEC archives)

"The opening of our school had been delayed about a month; for Habit School had been consolidated into the Philpot School district with a new brick three-story building—eight grades awaiting the coming of our electric power. Therefore, we went from a one-room eight-grade school with outhouses, water pump, and a potbelly stove to a new school of restrooms, water coolers, furnace, and lunch room.

"Yes, electricity gave country people all the advantages of our city friends. Bathrooms were built, appliances purchased, furnaces replaced grates and stoves, battery radios replaced with electric ones—and the store had drink coolers, a meat refrigerator, and fans.

"Yes, electricity and paved roads were a wonderful improvement in our lives. We all declared that first day of 'power' a holiday!!

"Note: after selling the store in June 1950, my father, Miller Taylor, went on to be employed by Big Rivers—securing rights-of-way—for *more* power lines!!"

KENERGY CORP.

Miles of Line:	6,944
Consumers billed:	54,095
Wholesale Power Supplier:	Big Rivers Electric Corporation/ Louisville Gas & Electric
Counties Served:	Breckinridge, Hancock, Lyon, Union, Caldwell, Henderson, McLean, Webster, Crittenden, Hopkins, Muhlenberg, Daviess, Livingston, and Ohio

ADMINISTRATION

Dean Stanley	1999 – 2004
Mark Bailey	2004 – 2007
Sandy Novick	2007 – present

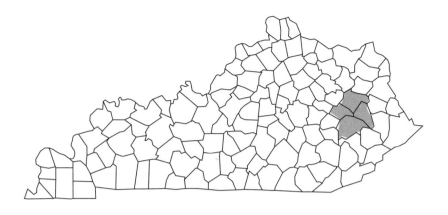

LICKING VALLEY
RURAL ELECTRIC COOPERATIVE
CORPORATION

If we had no electricity we would be back in the dark ages.

Kerry Howard
General Manager & CEO
Licking Valley RECC

Kerry Howard is mindful of the needs of the area he serves, so he usually doesn't wear a suit and tie at the office. A visit to his unpretentious headquarters at 271 Main Street in West Liberty is a reminder that places like Morgan County are family-based; they don't regularly go around putting on airs. The people of the heart of the mid-eastern Kentucky mountains—Breathitt, Elliott, Lee, Magoffin, Menifee, Morgan, Rowan, and Wolfe counties—are generally more comfortable with CEOs who don't lose touch with their upbringing. And the same might be said of all Kentucky co-op leaders. Their life's work is built on serving the electrical needs of the people, just as J.K. Smith firmly urged.

The annual Sorghum Festival in West Liberty attracts thousands of visitors from many parts of the country, but it's important to remember how the rural areas—Pleasant Run, Big Sinking Creek, and Toms Branch—waited so long for something as basic as a light bulb, something as comforting and reassuring as indoor plumbing.

Kerry Howard rises from his desk and moves politely aside to make room for Harry Steele, a retired fifty-five-year employee at Licking Valley RECC. Harry looks back at the time before the lights went on along the headwaters of the 320-mile Licking River—the main stream originating with Sprucepine Fork, Straight Fork, and Ann Cave Creek in Magoffin County. Licking flows on northwesterly through Morgan and Wolfe counties, once a wilderness area left in the dark by decades of unavailable electricity. For youth like Harry Steele (he's seventy-three now), it was a rugged struggle for survival and a long way downstream to the mouth of Licking across the river from Cincinnati, a brightly-lighted metropolis with its dazzling skyline.

"I was just a little boy when the house was wired, and we got electricity. And they were so much brighter than that old coal oil lamp. There were some houses that weren't electrified, and when I'd go visit, the light seemed so much dimmer.

"Our water well was twenty-five feet deep with fifteen feet down to water. If you turned over the milk bucket cooling in the well you had to pump the well dry."

Harry Steele was born in Morgan County about four and a half miles south of West Liberty. That's over in the Lykins Branch neighborhood where it took a "good" Jeep or a sure-footed mule to get back up most unlit hollers along streams like Coal Fork, Turkey Branch, and Honey Branch. It was decades before I-64 and the Bert T. Combs Mountain Parkway brought four-lane paths to the land of lights, opportunities to break free from the darkness of isolation to the openness of new possibilities.

"My grandfather had a springhouse—built their house over the spring and that's how they kept their things chilled."

The year was 1934, long before the lights went on, a full year before President Franklin Delano Roosevelt took pen in hand and created the Rural Electrification Administration. Something must have told him there was an ugly problem to be solved, an omission, a default not

When three stout-hearted men and a co-op truck are a team, it's a good bet that thousands upon thousands of very pleased men, women, and children won't be in the dark. (Licking Valley archives)

making any kind of moral sense whatsoever. What justification was there to leave about ninety percent of the rural areas of a great nation as dark as the inside of a cave? For every electrified town and urban area there were countless good, God-fearing farmers. Electricity was a miracle, but it was not fully shared.

The decade of the 1930s witnessed catastrophic drought in Oklahoma, brutal backdrop for John Steinbeck's *Grapes of Wrath*. Farm prices had fallen more than fifty percent, and foreclosures blotted the landscape. Many Midwest farm families gave up and headed for California, while rural Kentuckians, like the Steeles of Morgan County, held fast and made do before the lights finally went on, sometimes as late as the '40s and even the '50s. Of course, there was migration to the factories of Ohio and Michigan, urban areas that had all the electricity they needed from private power companies, but the trips back to Appalachia were stern reminders of the darkness of home.

Stringing the lines of Licking Valley RECC was tedious work. But when the lights first came on it was as if a new world had dawned. It would no longer be an unlighted land, isolated and forbidding. There would be a sign of hope.

"At first [in the early '40s], we had a ceiling light for each room. It was so much brighter than the old coal oil lamp. The refrigerator was the first appliance. We stayed with the old washing machine. I helped set poles for security lights," says Harry, who toiled a lifetime for the co-op made possible by REA funding, local grit, and a unified community view of the future.

Harry's father, Estill, was a farmer, and Mrs. Steele was a schoolteacher. She started in a one-room school until fourth or fifth grade. "It was a wonder how anybody could see to read," says Harry.

Harry, working thirty years as a pole setter for Licking Valley RECC, says he "likes a job, not a position." He's a seasoned Kentucky worker, and for the past twenty-five years he's been helping with special co-op projects.

Harry has been married fifty-five years. He and his wife Carleen have seven children, sixteen grandchildren, and one great-grandchild. It's safe to say, future generations of Steeles will be involved in keeping the lights

Remember working the old hand-operated bucket crank on the well? Remember not knocking into the butter and the milk hanging in the well while bringing up the bucket? No? Well, you weren't on a farm before the '50s. (KAEC archives)

burning to the farthest recesses of the eastern Kentucky mountains.

Then there's Jimmy. He too wants to tell his story, because he wants the city folks to know how it really was, not so very long ago.

Jimmy Adkins was ten years old when electricity "came up the road." Jimmy retired in 2001 after twenty-five years with Licking Valley RECC's public relations office. The earlier time was in the field, cutting right-of-way. He stayed with the co-op for forty-four years, give or take.

Before "Let there be light" leaped off the page of the old family

The look on the face of this sweet young lady is one of "I can't believe this big, ole bike is mine!" as she poses with the prize she just won at a Licking Valley RECC annual meeting. (Licking Valley archives)

Bible and became something more than heavenly words—before lives were turned around for five girls and four boys in the Adkins family—it was necessary to draw water for the weekly Saturday night bath. Water was drawn and heated outside in tubs, which were then brought inside the house—set up the tub in the yard, fill it about halfway to the top, manhandle it into the kitchen, then reach for the lye soap. "Make you red as an Indian," Jimmy remembers with a smile, a sparkle in his eye. He's been a patient, long-suffering Appalachian taking pride in sense of place.

Baths were taken by the youngest children first, and it might take four or five tubs of water for the nine children. Girls bathed in a room separated by a curtain from the boys. "You watched your mouth," says Jimmy.

In 2008, Kerry Howard is General Manager of Licking Valley RECC, responsible for parts of Breathitt, Elliott, Lee, Magoffin, Menifee, Morgan, Rowan counties and all of Wolfe County—more than 2,000 miles of line to more than 17,000 customers. Forty-four employees keep the lights turned on where once there was darkness after the sun descended toward the bright lights of Lexington and Louisville.

The cooperative was incorporated in 1940 with four hundred and thirty-one owner-members. The one hundred and eighty-nine miles of original line were first energized in May 1941.

We turn to the dispute about "mountaintop removal." How can it be defended? Should not the mountains of Appalachia be left undefiled? Should not the hollers ring with the music of song birds, banjo pickers, and the soft words of romantic tales, maybe a feud or two, instead of armies of behemoth bulldozers and monster earthmoving augers?

"If there was no electricity, we'd be back in the dark ages," says Kerry Howard. "If we didn't have mountaintop removal, the economic impact would be bad. We live off the sale of coal." Even "wheelage" has entered the language, taking its place with "wheel and axle" and "wheel and deal"—leasing arrangements in which property owners are paid for just the access to the sites of coal. "Glasgow [in Barren County in south central Kentucky] has beautiful rolling land, but we use the edge of the creek for a place to live. Mountaintop removal means airports, elk herds, and jobs at Wal-Mart. The land belongs to the people."

Our Power Is Our People is a framed print on the corridor wall leading to the office of Kerry Howard, leader of an organization proud of its more than forty employees, many scholarships, and the ongoing Health/Wellness program.

Who are some of the other beneficiaries? Maudie Nickell has been a Licking Valley secretary for thirty years, and Ruby Easterling has won a quilt-making show—she's ninety-seven years old. You wouldn't be far wrong in saying she's among the quiltingest—the art of quilting young

Taking care of downed lines in a flood or reading meters when the water is high is all in a day's work for Herman Arnett and General Manager Edd Gevedon. Doesn't hurt to know how to handle an oar in addition to an extendo stick to turn the transformers back on. (Licking Valley archives)

in her heart. True, over decades of troubled years many have left the mountains to seek more sophisticated, better-lighted lives, but there's a rich vein of Appalachian tradition built squarely on a willingness of neighbors helping neighbors—quilting bees representing cooperation within the community. And the men, they too have recognized the necessity to work together to transform what had been for so long, so dark.

One young man looks out the window of the pickup truck he's driving to the outer reaches of Morgan County; he remembers how his father worked in the mines—"He brought out coal with a pony."

Mountaintop removal?

"Magnificent if properly done and handled. God gave us this coal in these mountains. He left it to us to get it out and do something with it."

And what about all those who campaign to bring a halt to mountaintop removal?

Sometimes the prizes at the annual co-op meetings are like gifts from above. A brand-new deep freeze or hot water heater can make a huge difference when it comes to putting food on the table or hot water in the tub. (Licking Valley archives)

"You're not going to tell me what I'm going to do with my hill."

And so the mountaintop removal issue is passionately argued. On one side are those who see it as an environmental calamity, while there are others who see it as "my hill" with a God-given right to use it as I the owner see fit. Kentucky authors Wendell Berry and Silas House are leading the opposition to mountaintop removal. The outcome rests in the hands of the state legislature, on local politicians, and the voters. The people will decide.

The Licking River flows on.

Greater blessings couldn't come to the patients of the late Dr. Louise Caudill of Rowan County. In 1948 Nurse Susie Halbleib of Louisville joined her and for fifty years they made their way by jeep, horseback, sled, wagon, and by foot up the hollers of Rowan, Carter, Bath, and Fleming counties. Susie and Louise brought their own lamps to find their way to kitchens and bedrooms where the only light was the lantern. They worked in the tradition of Mary Breckinridge's Frontier Nursing Service, going where they believed they needed to go, doing what they believed they needed to do. It might be a broken arm or leg; or it could be something as simple as hard wax in an ear. No malady would be minor; sometimes, there might be a life-threatening situation in a cabin where electric lights were bathed in dreams of a distant future.

Louise and Susie delivered 8,000 babies, many in unlighted homes, where "we didn't see very well," Susie vividly remembers. And there was no running water, but "nature takes care of these things"—except when there might be lacerations.

Susie quietly recalls the time when a diuretic was urgently needed for a heart patient. The woman might have lived longer if a vein could have been seen.

There will always be storms, outages, and bothersome meter readings, but they are far better than no electrical power at all. Total rural electrification was long overdue but it finally became a reality as the result of cooperative effort and inspired leadership.

J.K. Smith foresaw the benefits that would one day come. As he said, "...if we work together in mutual understanding for the common good, then we need have no doubts or fears of the future."

LICKING VALLEY RURAL ELECTRIC COOPERATIVE CORPORATION

Miles of Line:	2,018
Consumers billed:	17,266
Wholesale Power Supplier:	East Kentucky Power
Counties Served:	Breathitt, Elliott, Lee, Magoffin, Menifee, Morgan, Rowan, and Wolfe

ADMINISTRATION

Kelse H. Risner	1940 – 1944
Fred Rose	1944 – 1954
W.E. Gevedon	1954 – 1982
James W. Elam	1982 – 1984
Bill Duncan	1984 – 2006
Kerry K. Howard	2006 – present

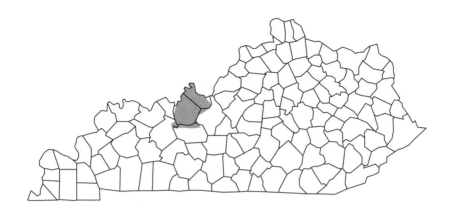

MEADE COUNTY
RURAL ELECTRIC COOPERATIVE
CORPORATION

A co-op is here for one reason and
one reason only, to serve our members.

Burns Mercer
President and CEO
Meade County RECC

Winifred "Tootsie" Hedden was born the same year Meade County
RECC took its first breath, the same year President Franklin D.
Roosevelt began his second term with the stirring battle cry: "I see one-
third of a nation ill-housed, ill-clad, ill-nourished."

The worst was yet to come. One week after F.D.R.'s second inaugural
address, the Ohio Valley flood of 1937 rolled in a mighty torrent from
Ashland, Covington, Louisville, Owensboro, Henderson, and Paducah to
the Gulf of Mexico. Millions were left homeless and without livelihood.

In the decade of the 1930s, Kentucky tenant farm families might be
inundated by flood, might be drawn up by drought, but the underlying
damage was the virtual nonexistence of rural electric power. The cities
had plenty of it, even enough for night major league baseball games, but

it was a different story in rural areas like Meade County's neighborhoods along Wolf Creek, Sirocco, and Hog Wallow.

For families like Tootsie Hedden's mother and father and their nine children, it was a test of strong character, grim determination, and pioneer values.

"Times were pretty rough. We didn't go anywhere, didn't own a car," says Tootsie. Home was a log house on KY 44, still standing about five miles west of Taylorsville in Spencer County. The original homeplace faces KY 623 near the spot where Plum Creek empties into Salt River. Conditions just over the hills repeated in Meade, Breckinridge, and Hardin.

"Mom cooked on an old wood stove. She kept kindling in the firebox for slow cooking. She kept old, heavy irons, six or eight of 'em on the stovetop for pressing clothes."

Tootsie had four brothers and four sisters (one died early). "We all had a job to do. We had coal oil lamps, so we went to bed early and got up early. We were not allowed to not do anything. We worked all the time.

"We had an old icebox that held big lumps of ice. There was a springhouse for drinking water, and we set milk and cream in the cold water flowing through. We grew tobacco and corn, raised cattle, hogs, sheep, and chickens. We raised our own beef to kill. We were typical country farmers — tenants, but later we owned."

Then a miracle turned dark into day. Or so it seemed. On the other hand, it might be easy not to miss something you didn't have in the first place. But as Grandmother used to say, "What's good for the goose ought to be good for the gander," which means in cooperative context, what's good for city ought to be good for country.

"Father didn't talk too much, but when the co-op started, he was a part of it," says Tootsie. "When our house was electrified, they ran one wire to the front porch. Then it ran across the ceiling to the place where you needed it. When you got a new appliance, you had to call the electric company so they could send more power to your house. After electricity came, everything changed. People began to do better for themselves." The gander was about to catch up with the goose.

The Board of Directors sets the tone of the cooperative movement and accepts responsibility for all outcomes. Those who served in the early '70s were: (L-R) (back row) Gary Pile, Jim Sills, J.D. Cooper, and Beavin Thornsberry Jr., Manager; (middle row) David Wilson, Bill Seaton, Leroy Humphrey, J.R. Watts, Attorney; (seated) Ray Barton, Joe Hamilton, and John Burnett. (Meade Co. RECC archives)

Leslie Jenkins offered a brief historical glimpse of Meade County RECC as told at a Public Service Commission hearing.

"Upon assuming my duties as Manager of the Meade County RECC in 1951, I immediately began to make a study to determine the needs of the cooperative.

"The cooperative was formed in 1937 and began operations in December 1938. At the beginning, one substation was owned by the cooperative at Tip Top, Kentucky, and all members were served through a distribution line. By 1950, due to increased load and LG&E's refusal to do so, it was necessary for the cooperative to construct approximately forty-seven miles of transmission line and add three additional substations to meet the increased demands for electric power.

"After completing the study and analysis of the cooperative in late 1951, and early 1952, it was evident that if the cooperative was to

The early office staff in front of the early Meade County RECC office. These ladies kept the records, collected the meter fees, paid the bills, and generally kept the office humming for the member/owners. (Meade Co. RECC archives)

survive, and to deliver adequate low-cost power to its members, several problems needed immediate attention.

"One major item was the wholesale power cost to the cooperative. Another was voltage regulation. Additional substations and transmission lines were needed to carry the extra load and correct regulation.

"Although Meade County was a member of the East Kentucky G&T Cooperative, there was no relief in sight for us until about 1958 or 1959, since the original loan to East Kentucky did not include funds to construct the necessary facilities to serve Meade County RECC. I immediately set up negotiations to see what could be done about cheaper wholesale power, installing additional substations, and building additional transmission lines to correct voltage conditions.

"My first negotiations were by telephone conversation. One was with Powell Taylor, engineer, of Kentucky Utilities Company, now deceased, requesting a connection on the west end of the Meade County RECC's system to relieve the voltage deficiency and otherwise firm up the transmission and distribution system. I talked with Louisville Gas and Electric Company in regards to this matter and received their consent to contract with KU. Before the negotiations were completed, Kentucky Utilities advised us they would not serve the west end of our system.

"I immediately went back to Louisville Gas and Electric Company for further negotiations. After several meetings we were successful in negotiating a five-year contract with Louisville Gas and Electric Company for wholesale power. This contract was negotiated on September 30, 1954.

"From 1951 up to the present time, the cooperative has made tremendous investments, not only in distribution lines but in additional transmission facilities and substations. It now owns and operates over eighty-two miles of transmission lines and seven substations. At the present time, an additional fifteen miles of transmission and one new substation have been approved by this Commission and are now under construction. As I stated previously this eighty-two miles of line cost us $615,461."

"Did LG&E deliver their power to your required load centers?"

"Absolutely not. They always required us to construct our lines to their existing facilities."

"Were you ever able to get another point of delivery from them?"

"Yes. On September 29, 1956, we had to sign a ten-year contract in order to get one new delivery point. This is our contract now in force, expiring on September 29, 1958."

"Were you satisfied with the ten-year contract?"

"Of course not, but we were forced to sign this contract in order to get the new delivery point, which we had to have to meet our minimum system requirements. We ordinarily would have only agreed to a contract of five years or less. Since we were participating in negotiations for a new source of power supply to relieve us of our wholesale power contract restrictions, we wanted to sign a short-term contract. Our immediate need for this new delivery point forced the ten-year contract on us. Now we cannot buy power from Big Rivers until 1969. Therefore you can see why a long-term contract is not flexible enough."

"Why were you interested in a different source of power supply?"

"We knew it was the only way we could get a better wholesale rate, and get out from under the restrictions in our contract with LG&E."

"What then did you do?"

"As I said, Green River RECC, Henderson-Union RECC, and Meade County RECC joined together."

The words "joined together" resonate in the world of rural electric cooperatives. "Join" embraces the idea of positive movement toward unity, setting aside selfish motives; "together" results in concerted strength that enables individuals to accomplish more through teamwork. Still, there have been obstacles.

Current President and CEO Burns Mercer writes: "It seems every Manager/CEO at Meade County RECC has encountered unusual and/or challenging events during their tenure.

"Curtis Brown (1939-1951) had the difficult task of basically starting a new business. Les Jenkins' primary task when he arrived in 1951 was to turn Meade County RECC into a more professionally run co-op. Upon Les Jenkins' retirement in 1974, Meade County RECC was indeed a more professionally run co-op with a stable cooperative-owned power supply. The Board then elevated a longtime employee, Beavin Thornsberry Jr., into the position of Manager. Mr. Thornsberry was immediately thrown into the middle of a labor dispute with the line workers.

"During this time period the tornado of April 3, 1974, struck Brandenburg. In addition to the thirty-one lives lost, the tornado did major damage to the lines of Meade County RECC as well as completely leveling our headquarters building. All employees escaped to the basement seconds before the tornado hit and no co-op employee lives were lost.

"Operating out of temporary headquarters, repairs were eventually made. Also, twenty acres were purchased at the edge of Brandenburg and a new headquarters was constructed and moved into in the fall of 1975.

"Toward the end of Mr. Thornsberry's tenure, a sad chapter was unfolding in the history of our power supplier Big Rivers.

"Throughout these times Meade County RECC saw steady growth. Meters billed went from 12,896 in 1974 to 20,531 in 1994, assets rose from $8,600,000 in 1974 to $31,000,000 in 1994, and revenue increased from $2,500,000 in 1974 to $19,000,000 in 1994.

"Technological changes and power supply opportunities and challenges have been the theme so far during my tenure as CEO.

"Due to Big Rivers' difficulties, the Big Rivers and Meade Boards and I were confronted with a major problem [how to right the power supply ship]. The immediate solution was to form a committee [the work-out committee] to decide what course on which to proceed. As a member of that committee [I] and other committee members came to the conclusion that a bankruptcy of Big Rivers was the only solution to confront its mounting problems.

Four horsemen of a surveying team that not only use a transit to accurately determine position points and distances between them but a machete to whack at the weeds in the way of the sightings. The building of power lines requires well-marked paths and legal records. (Meade Co. RECC archives)

The building of high-voltage stations demands agility, carefulness, and respect for hazards. These two Meade Co. RECC men seem not only to know what they're doing, but do their jobs well and in good humor. (Meade Co. RECC archives)

"After many months of negotiations and court proceedings, Big Rivers entered and then exited from bankruptcy in 1998, a much different organization.

"Things went well for Big Rivers and Meade County RECC from 1998 to 2004. In 2004, LG&E notified Big Rivers they were not satisfied with lease arrangements and in fact wanted out of that agreement.

"[I] was appointed to a task force [as the members' representative] to try and negotiate an end to the lease and a resumption of Big Rivers control over the power plants.

"This process is now over four years in the making and Big Rivers, as well as the member co-ops and the other parties to the transaction [LG&E and the aluminum smelters] are currently before the Kentucky Public Service Commission and expect an order soon and a closing of the transaction sometime during the summer of 2008.

"This closing should again put Big Rivers and the member co-ops back in control of their power supply once again.

"Since 1994, technological change has been a constant at Meade County RECC. A new GPS mapping system, automated meter reading, etc., are just a few of the technology changes that have been implemented and no doubt will continue to be implemented in the future.

"Also, during this time Meade County RECC continues its growth since 1994. Meters billed have grown from 20,531 to over 27,500, assets have increased from $31,000,000 to over $71,000,000, and revenue has increased from $19,000,000 to almost $30,000,000 in 2007.

"This story of Meade County RECC has been told from the perspective of its Manager/CEO, but in fact all of its dedicated employees and directors are responsible for the success of the co-op."

J.K. Smith's words ring true: "We have to have some vision about where we're going and how we're going to get there. It all comes down to common sense: observing, being up to date, staying on top of what's going on, being aware, anticipating problems down the road and trying to plan for them. If you use common sense, you can do just about anything."

Six early linemen provide the balance needed to offset the 6,600-pound transformer being positioned on the back of a flatbed truck in the late 1930s. At the time these transformers were installed, they cost $2,000 apiece. (Meade Co. RECC archives)

Or as simply stated by the people—owner-users like "Tootsie" Hedden—"After electricity came, everything changed. People began to do better for themselves."

MEADE COUNTY
RURAL ELECTRIC COOPERATIVE CORPORATION

Miles of Line:	2,948
Consumers billed:	27,459
Wholesale Power Supplier:	Big Rivers Electric Corporation
Counties Served:	Breckinridge, Grayson, Hancock, Hardin, Meade, and Ohio

ADMINISTRATION

Curtis Brown	1939 – 1951
Les Jenkins	1951 – 1974
Beavin Thornsberry	1974 – 1994
Burns Mercer	1994 – Present

Note: Not listed are two managers who served briefly in first two years.

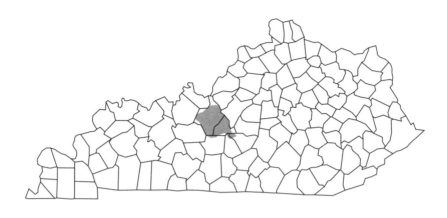

NOLIN
RURAL ELECTRIC COOPERATIVE
CORPORATION

The dream that had begun on that warm June day
in 1937 was finally realized on July 1, 1939,
as the first section of line was energized.

Michael L. (Mickey) Miller
President & CEO
Nolin Rural Electric
Cooperative Corporation

By 1938, more than a dozen rural electric projects had been organized in Kentucky, where only three percent of the farms had electricity. One such project was started that year in Elizabethtown. Pauline Duff, an attorney from Sonora, was a leader in that effort. What follows is her account of the early days of the Nolin Rural Electric Cooperative Corporation as it appeared in the *Hardin County Enterprise* on Thursday, June 29, 1939.

"While credit is being given the Board of Directors for the present status of our Cooperative our project has succeeded mostly through the loyal and untiring efforts of countless neighbors and friends who have gone down the road and sold electricity to the doubting.

"A warm day early in June, 1937, Mr. A.J. Thaxton, county agent, called a mass meeting at the courthouse of all interested in electricity. Approximately one hundred and twenty-five were present, and at this meeting committeemen were appointed, one from each of the six magisterial districts, who with helpers were to make a canvass of the territory to find out the number of people interested in electricity. On August 20, 1937, there was another mass meeting at the courthouse when a temporary board of directors was nominated, and afterward the board met and selected its officers. Canvass sheets together with a map showing the location of each farm home were sent to Washington in October, 1937. However, as funds for 1937 had been exhausted, we had to wait until the fiscal year beginning July 1, 1938, when an extra $100,000,000 was appropriated by Congress.

"On June 13, 1938, Mr. Russell Cook from REA in Washington arrived unexpectedly to look over our proposed project. He gave us his OK, and Hardin County was immediately given permission by REA to begin a project.

"Therefore, on Friday night July 15, 1938, the temporary board now composed of five directors—Mr. W.R. Crawley, Mr. Harry Gatton, Mr. J.C. Brown, Mr. Grover Johnson and myself—met in Mr. Thaxton's office and appointed Mr. H.C. Smith of Brandenburg and Mr. H.L. James, Jr., of Elizabethtown as project attorneys to immediately draw up incorporation papers in the name of the Nolin Rural Electric Cooperative Corporation for the signature of the Board and to be filed immediately with the Secretary of State. Our charter was granted July 20, 1938, and on July 21, 1938, followed the necessary meeting of the Board as directed by REA when the packet pertaining to 'Pre-allotment Procedure for New Projects' and 'Instructions for Obtaining Memberships and Easements' were read by the Board and fully explained by Mr. James preparatory to taking of memberships and easements. This work began the following day.

"It might prove interesting to know why the name 'Nolin' was chosen—it was originally intended to have our Cooperative named 'Nolin River,' but as the name was so long, it was decided to drop

They come to the annual meetings, always have, with determination and resolve that their company is doing the best job possible. (Nolin RECC archives)

'River.' This name was the Board's selection after much discussion due to the fact that we were advised that our project would cover part of LaRue, Hardin and possibly Grayson Counties, and the Directors felt it would be for the best interest of the Cooperative to have a name that would not especially identify the project with Hardin County alone.

"One month after our membership canvass started and we had reached a membership nearing five hundred, a committee from LaRue County composed of Mr. F.G. Melton, county agent; Mr. Roy Ragland, Mr. Thos. Ovesen and Mr. Bob Meers attended a survey leaders' meeting and afterwards met with the Board when an invitation was extended to LaRue County to come in with Hardin County on the original survey. This was accepted and a separate canvassing crew was chosen to solicit LaRue County. In the summer of 1937, LaRue County, like Hardin, under the ably organized LaRue County Farm Bureau, canvassed for prospective members but was turned down due to the fact that while they had the necessary members to a mile, they did not have sufficient mileage to warrant a separate project.

"On August 21, 1938, there was a statewide REA meeting in Louisville at the Brown Hotel, and the entire Board attended. We heard REA officials and engineers from Washington explain in detail the ideas and purposes of rural electricity and, after the general meeting, the Hardin County group had a private conference with the officials there.

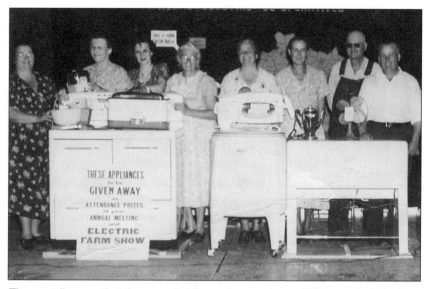

These smiles say that the new electric appliance prizes will fit very nicely in the old kitchens where once stood the wood-burning stoves and the #3 wash tubs. What a welcome relief from drudgery! (Nolin RECC archives)

From this time on, more progress was made.

"On September 16, 1938, the pre-allotment survey was begun by Ray W. Chanaberry, Inc., engineers of Louisville, Kentucky. One month later, this was completed and all necessary maps and papers were signed by the Board and were sent to Washington by Mr. Chanaberry.

"On December 10, 1938, a telegram was received from REA in Washington stating $328,000 had been allotted for this project consisting of 301 miles in Hardin and LaRue Counties. Immediately Ray W. Chanaberry, Inc. drew up the necessary detailed information to permit bids to be received on the construction of the lines. On December 21, 1938, bids were opened at the office of the Cooperative and the contract for construction was given to R.H. Bouligny, Inc., Charlotte, N.C., who was the low bidder.

"One week later, Mr. Parker Dodge and his assistant and surveying crews were sent here by REA to begin the actual surveying and staking of the lines. On January 6, 1939, there was an all-day meeting of the board. Thereupon in the name of the Cooperative, mortgage and note was executed as directed by REA in the sum of $328,000 to secure our loan. The principal payment on the note does not start until thirty-one months after the execution of the note and mortgage, and REA has set up a budget for our project wherein the interest and principal payments are to be taken care of out of the earnings of the Cooperative through the payment of the monthly light bill. Due to the many delays there were in getting actual construction under way, which held up the honoring of requisition for funds, it was necessary in February and March [1939] in order to keep the work going to have funds with which to meet the weekly payroll of the engineers, etc. The directors personally borrowed $1,000, as the Cooperative is forbidden to borrow funds. This was paid as soon as funds were received as allowed by REA.

"Mr. Crawley was acting superintendent until March 12, 1938, when Mr. Frank L. Hagan received the OK of REA in Washington for the position of project superintendent.

"Early in May, Mr. Parker Dodge was transferred to another project at his home in Flemingsburg since almost all survey work had been completed and released to the contractor. Mr. F.M. Slater was appointed as REA local engineer for our project, his assistant being Mr. Danzel Murphy. These two with Mr. Jimmy Broaddus, REA district engineer, have full charge of the final checking up of our project and seeing that it is built according to specifications and requirements of REA.

"On June 13, 1939, Mr. Jimmy Lyons of Hodgenville was appointed as maintenance engineer and is now busy installing meters and connection service.

"Everything seems to now be running smoothly and according to schedule with most right of way trouble being adjusted without resorting to lawsuits. We are now on the eve of our energizing celebration—in other words, we are ready to give thanks and to rejoice that once again Hardin and LaRue Counties have passed another milestone along the road to success.

"This would not have been possible without the cooperation of many, many who do not live on their farms and to these many we say, 'Thank you,' for you have been GOOD NEIGHBORS in giving right of way easements across your farms and, in a number of instances, through a sense of public spirit have even signed up your farms. In doing these things you have made it possible for the country to have all the conveniences of the city."

From an article edited by President and CEO Mickey Miller in the Nolin RECC archives, comes: "Before rural electrification, life in rural America essentially began at sunrise and ended at sundown.

"Electric power was one of the key attractions of city life in those days

In the beginning, meters were read by the user/owners, then came meter readers. Now, thousands of meters are automatically read in the home office. How times have changed. Hooray! No more dog bites for the meter readers! (Nolin RECC archives)

and contributed to the exodus of many rural residents, especially young people, from the farms.

"Country folk would make the best of it with kerosene lanterns, washtubs, scrub boards, wood stoves, flat irons and outdoor toilets (sometimes 'three-holers,' but more often 'one-holers'). Rural economies were tied almost exclusively to farming, and rural life in general consisted of the same toilsome routine that had existed since colonial times.

"If you were a farmer, you began your day walking by lantern light to the barn where you milked cows by hand. You ended it the same way. You hauled water and mixed feed for the animals by hand because you had no electric pump or grinder. You hoped where you were in the barn your lantern would not overturn accidentally and start a fire.

"Life was no better if you were a farm wife. The sheer drudgery of hauling wood and water left you stooped and bent from an early age.

"Your days also began before sunup. Starting a fire in a wood stove and keeping it going was no easy task if there was a draft in the room or the wood was wet. The stove furnished warmth in the winter, but in the summer, its heat made the house, especially the kitchen, an inferno. Whether to cook a meal, heat an iron or boil water for washing or canning, you had to keep the stove fired up all day and its heat remained long into the night.

"You preserved your food in a blockhouse, a smokehouse or a cellar. If you lived close to town and could afford it, you had an ice box for storing a few perishables and an ice man who delivered large blocks two or three times a week. If you didn't have an ice box, you kept your milk, butter and fresh produce in a springhouse or a large bucket at the bottom of a well. Neither was very convenient. With a well, you had to remove the food items every time you needed to draw water. With a springhouse, it was not uncommon to find an overnight flood had carried your food downstream or a warm spring rain had raised the temperature inside the springhouse, causing the food to spoil."

Setting a pole is a test of endurance, strength, and experience. It all had to be done the hard way; holes dug by hand, and poles set by hand with the aid of a truck and winch. (Nolin RECC archives)

195

Nolin RECC member Bob Owsley remembers February of 1939, when electricity came to the Owsley farm: "The old wood burning cook stove in the kitchen was later disposed and replaced with a modern stove…such a wonderful extravagance. After those early years, an electric mixer, iron skillet, toaster, radio, and many other items were acquired. Finally, other plug-ins were installed for an electric washer. A deep well was drilled outside and an electric pump was needed."

Diane Skaggs Osborne of Round Top Road in Hardin County shares this: "My daddy, John Durham 'J.D.' Skaggs, and my mom, Rachel Sidebottom Skaggs, married April 3, 1947, and started their life together on a 140-acre farm seven miles from Hodgenville.

"My precious mother must have really loved my daddy to have waited for him to return from WW II and then to move from her parents' home in Hodgenville, which was warm, comfortable, inside bathroom, running water, and electricity, to a farm into an old farm house with no inside plumbing, kerosene burning stove, and worst of all NO electricity.

"The actual day it [electricity] was to run into their home, my mother and I were in Hodgenville visiting my mother's parents. Daddy said he could hardly wait for our return. Finally he saw us turn into the drive. He ran out to greet us and to quickly steer us into that old home. He said, 'Watch this, girls,' as he very happily pulled the cord that immediately turned on the one and only bulb he had in our home. It shone so brightly! He said that was one of the happiest and proudest days of his life!"

Willis Willyard, a man who never spent a night away from home but once (he lost two fingers in a corn picker and had to stay overnight in a hospital), grew up on the Rineyville-Big Springs Road in Hardin County. He and his parents didn't get electricity until 1939. He was twenty-one years old at the time.

"We didn't know any better."

"You were active in the creation of Nolin RECC?"

"Dad was more active. There had to be three users to the mile. Our next door neighbor wouldn't take it. Some people couldn't accept all that newness. But the lines were brought in and the neighbor decided to take it."

"Do you miss the farm?" Willis is ninety years old this year; he and

A man-to-man meeting, with a little youth on the side. Co-op Manager Clem S. Tharp stayed in touch with members in his service area in 1964—those like Mr. Lee Cundiff and his granddaughter in LaRue County. United we stand, divided we fall. (Nolin RECC archives)

his wife Lottie have moved to Elizabethtown, one of those early electrified places.

"Time has crawled up on us," says Willis, who remembers helping his father milk by hand twelve to fifteen cows every morning and night. He remembers how his mother cooked with wood cut for that purpose, while other wood was cut to keep the house warm. Water? It came from a cistern.

Life was anything but easy for Willis Willyard until "let there be light" became a cherished reality.

NOLIN RURAL ELECTRIC
COOPERATIVE CORPORATION

Miles of Line:	2,900
Consumers billed:	31,488
Wholesale Power Supplier:	East Kentucky Power
Counties Served:	Breckinridge, Bullitt, Grayson, Green, Hardin, Hart, LaRue, Meade, and Taylor

ADMINISTRATION

W.R. Crawley	1938 – 1939
Frank Hagan	1939 – 1943
C.D. Thompson	1943 – 1945
T.A. Heywood	1945
James A. Lyons	1945 – 1958
Clem S. Tharp	1958 – 1981
Jack H. Kargle	1981 – 1991
Michael L. Miller	1991 – Present

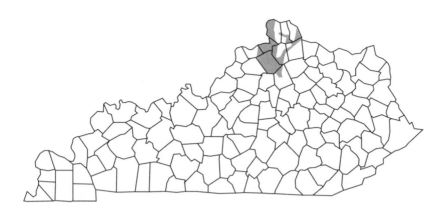

OWEN ELECTRIC
COOPERATIVE

The importance of service to the membership
will continue to be the driving force
behind all of Owen Electric's efforts.

Bob Hood
President/CEO
Owen Electric Cooperative

"The possibility of having electricity on the farm was the topic of every conversation from 1935 until June 8, 1937 when Owen County RECC was organized and the Articles of Corporation adopted. [From *The Northern Light*, a publication of Owen RECC, September 1976].

"Charter members and the first board of directors were J.H. Satterwhite, President; Lister Ransdell, Vice President; Ira Kemper, Secretary & Treasurer; J.W. McElroy and J.L. Tackett. Chester Roland was hired as Project Superintendent July 10, 1937 at a salary of $125.00 per month plus mileage of not over $35.00 per month.

"The first annual meeting was held June 17, 1937, attended only by the five men mentioned above and fifty-two other interested countians.

Members of Boards of Directors live in the districts they represent and Directors are elected by the people they serve. In the 1960s, they were (L-R standing) A.W. Ayres, R.L. Penn, Roy Gray, W.M. Smith, Chester Campbell, Charles Richardson, General Manager, T.A. Perry, Attorney, (kneeling) Robert Arthur, Howard Greene, W. Oakley Noel, and Frank Jackson. (Owen Electric archives)

In two days they collected one hundred and twenty-four applications at $5.00 each. Bylaws were adopted at this meeting and a loan contract was signed with REA in the amount of $130,000 for construction of one hundred and thirty miles of line to serve approximately five hundred applicants in Owen County only.

"The first office was in the Fullove building with the rent being $5.00 per month. Judge James Cammack, Sr., was retained as legal counsel and remained in that position until his death, at which time H.W. Alexander succeeded him on March 3, 1939.

"Contractors were hired and almost eight months later on January 29, 1938, Owen County RECC energized its first lines and gave service to its first members. Governor A.B. 'Happy' Chandler was among the dignitaries who attended the celebration, his second such appearance following his first at Henderson in 1937.

"April 11, 1938, the board voted to include the neighboring counties of Grant and Gallatin to the service area followed by Pendleton County on June 27, 1938.

"A loan for $22,600 was signed and approved by the board on June 6, 1938. This loan was approved to extend service to additional applicants in Owen County. At the same board meeting J.L. Tackett resigned as director and was succeeded by D.F. Barker of Williamstown on March 3, 1939.

"March 4, 1939, the board voted to include Scott and Campbell Counties in the service area. They signed and approved another loan for $95,000 to again extend the lines.

"The lines were first energized in Grant, Gallatin and Pendleton Counties in January, 1939. In December of 1939 the members voted to include Boone, Kenton and Carroll Counties to the area. This brought the Cooperative to its present status of nine counties in the Northern part of Kentucky. Naturally more money was needed to extend these lines and another loan was signed and approved for $267,000. By the fall of 1940 some customers were receiving service in all nine counties. The Cooperative had grown by this time to approximately seven hundred miles of line and 1,965 members. World War II was about on us at this time and many rural people would have to wait for service because of the shortages created by the war. Owen County RECC was the largest Co-op in the State at this time.

"The Utilities were beginning to realize that the

Transformers step down voltage to make electrical power safely usable. There are acres and acres of transformers across Kentucky just waiting to be called into use.
(Owen Electric archives)

201

Co-ops meant business, and began constructing lines to rural areas. The Co-op had to refund membership fees to eleven people because Union Light, Heat and Power built to them after they had paid their membership fee.

"The largest consumer the Co-op had in 1941 was Twin Creek Mining Company. Their May bill was $90.00. The average bill for December was $3.36.

"The war had come to an end and the rural people were becoming even more eager for electricity. The country boys who had served their country had seen the bright lights and wanted them more than ever. The Cooperative had been investing its surplus in War Bonds during the war, and on February 6th cashed in a $10,000 bond to purchase line construction material.

"The annual meetings in those days were quite different than the ones held today. They were usually held in conjunction with the county fairs with no entertainment but displays of appliances by local dealers. The rural people were brand new consumers of all the electrical appliances that their city cousins had been using for years.

"The members agreed to increase the debt limit to five million dollars. It was also agreed to install two-way radios in the trucks to save mileage and time. In April, 1947 a central service billing contract was signed with East Kentucky RECC.

"Because of the growth in the northern area it was decided that a sub-office was needed. Five houses were leased in Butler, Kentucky, for a period of eight years. November of 1948 saw the beginning of Capital Credits.

"The Board of Directors in May, 1949, consisted of Roy Adams, Henry Heringer, Albert Davis, Lester Ransdell, Will Smith, J.L. Bentle, J.F. Burk, R.L. Penn, Stanley Parker, A.W. Ayres and Walter Luttrell.

"The same year an agreement was reached to buy our power from East Kentucky RECC, a Generating and Transmission Rural Electric organized in 1941, but unable to get off the ground because of the war.

"Owen County had been operated from rented office space, in what is now the First Farmers Bank building and Fullove building since it had been organized. It was announced in February of 1950 that the estimated construction cost of the new office had increased from $80,200 to $100,000.

"June 28, 1951 Owen County's new office and warehouse were completed. The building located on Highway 227 near Owenton served as the headquarters for the next 52 years for the Cooperative. When we moved into the new office the Cooperative had grown to 2,231 miles of lines for 7,243 members, or 3 1/4 members per mile.

"Because the rural people had not had the advantages of electricity and many of them hesitated to buy water heaters, electric ranges, refrigerators and other conveniences, it was deemed

The hard hat, the confident face, the well-gloved hands—all vital parts of getting the job done safely, efficiently, and effectively. Andy here is a good example of getting the job done right the first time. (Owen Electric archives)

necessary to promote them. It would not only help build a load and increase revenue for the Cooperative but would also relieve the work load of the consumer. The first was an electric range circuit promotion. Any consumer buying an electric range could make an application to the Cooperative and we would install a range circuit free. This was later changed to $25.00 then to $20.00, and in 1971, after a survey showing a high saturation of electric ranges on the system, it was dropped. The Cooperative had promotions on hot water heaters for $15.00, residential wiring changing from 60 amp entrance to 100 amp was $30.00 and from 60 amp to 200 amp was $50.00, central air condition was $75.00, electric heat was $75.00 and the Gold Medallion Award was $275.00."

Further issues of *The Northern Light* in November and December

1976 continued the story:

"Owen County's debt limit was increased to $15,000,000 this year and remained at that limit until a special members meeting was called January 19, 1972, and increased it to $30,000,000 so as to be in compliance with the new lending institution, CFC [National Rural Utilities Cooperative Finance Corporation].

"After a thorough study and cost analysis, it was decided to keep *The Northern Light* a separate publication rather than going to an insert with *The Rural Kentuckian.*

"1973 saw the end of the 2% REA money. Rising coal prices and underground services began to be an endless thing.

"In January of 1974 there was the longest sleet storm in history causing many outages. The cost of coal had risen from $5.00 per ton in 1971 to over $30.00 now and in some spot instances as high as $60.00 per ton.

"In December the interest rate with C.F.C. on short term loans rose to 10 1/2%.

"The 40th Annual Meeting was held June 29, 1976, at the Grant

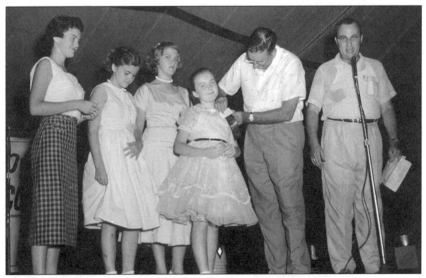

Recognition crosses the generations at the Owen County RECC annual meeting in 1957. Today's youth become tomorrow's users/owners and will remember fondly what it means to be a part of a cooperative. (Owen Electric archives)

County High School in Dry Ridge. The weather reports were for thunderstorms but the meeting proceeded as planned. The business meeting was held, and the first showing of the beauty contest. The door prizes were awarded with the exception of the boy's bicycle when the wind suddenly hit the tent with such force it was a miracle that it stood. The people ran from the tent to their cars, and to the high school gym. The crews managed to lower the outside edges of the tent preventing the wind from completely destroying it.

"Six people received injuries; the most serious was a broken collar bone.

"The Washington Youth Tour was started in 1972. Owen County invites the juniors in all the high schools of our service area to participate in the program. Two students are selected to tour the Capital each year with 1,000 other young people from all over the United States.

"Looking back over the past forty years we are proud of the accomplishments that have been made. There have been good years without incidents to cause much concern. Then there have been years that everything imaginable happened. Natural disasters, inflation, shortage of materials, high interest rates, coal cost going up over 300% and rate increases to name a few.

"It has not been an easy forty years, but through the support of you as member owners attending the annual meetings, paying your bills, reading your own meters, and with you unselfishly allowing us to go through your property to serve your neighbors we have become what we are."

President and CEO Robert A. Hood is asked what he believes are the three most significant developments in the history of Owen Electric:

"During 1937-1940, Owen Electric became Northern Kentucky's 'Pioneer' in furnishing electric service to the region. In 1937, a group of local citizens had the vision and courage to improve the quality of life in Owen County by organizing a small electric cooperative known as Owen County Rural Electric Cooperative Corporation. Soon, neighboring counties saw what was being accomplished and asked to join the cooperative. In 1939, Grant, Pendleton, Gallatin and Scott counties joined the cooperative. These nine counties still make up Owen's service area today.

"In 1990, Owen Electric returned over $1,000,000 in capital credits to its membership. In keeping with cooperative principles, Owen Electric's board and management announced the first general refund of capital credits to its members. This annual general refund continues uninterrupted since 1990.

"In 2001, Owen Electric continues its 'Pioneering' heritage by being the first electric utility in Kentucky to promote and sell 'green power.' Guided by a commitment to its members and the communities it serves, Owen Electric was successful in obtaining green power at the request of the U.S. headquarters for Toyota Automotive. From its humble beginnings in 2001, this initiative has grown into a significant part of the renewable energy production program of the East Kentucky Power Cooperative member systems. As of 2007, five landfill gas power generation plants are in operation in the state, producing over 15 megawatts of power.

"In November 2003, after fifty-two years in their offices on SR #227, Owen Electric moved into a new facility on SR #127 about nine miles north of the City of Owenton, which combined their corporate headquarters and their service center for the southern counties. This facility was designed to meet their needs for the next 50 to 100 years.

"Owen Electric continues modernization of its equipment, facilities, and technology. Owen today is recognized by the cooperatives of Kentucky as one of the most advanced in all of these areas. This modernization has always been driven as a way to provide more advanced and better service to our members."

And the future? How does Bob Hood see the years ahead?

"Owen Electric Cooperative will continue to be a leader and pioneer in the communities we serve. We will embrace dynamic and innovated approaches in providing quality service through new technologies and continual process improvement.

"Owen Electric's latest effort to elevate our service to a higher level is being achieved through our transition to an Advanced Metering Infrastructure (AMI). AMI will allow Owen real-time communication with our meters, leading to a host of service enhancements to our membership.

"The importance of service to the membership will continue to be the driving force behind all of Owen Electric's efforts."

A story from the RECC archives:

A lady living in Owen County always used exactly 40 kWh, which cost $2.75 a month. Month after month her reading of forty kWh and check for $2.75 would arrive on the due date. The maintenance man knew that the lady was honest, but he was curious as to how she could use the same amount each month. He stopped by one day and asked her how she managed. She very emphatically informed him that she watched the meter and when it got to forty she turned everything off.

Evidently her frugality paid off. She sent her son through medical school and he has been a doctor in Carrollton for years.

Betty Shelton Laurence was raised in southern Owen County in the New Columbus neighborhood: "Life was so dark at night and until electricity came through, you could see a light on every hill where so many people lived. Mother washed on a tub and board. She'd put back so many of Daddy's overalls where he had sheared sheep and cut

Next time a traveler is up "Sweet Owen" way, a visit to the Owen Electric Cooperative's museum at the main headquarters on Highway 127 N. is more than worth the effort. (Photo by Whitney Prather)

tobacco—when we got our first wringer washer, Daddy had enough clean overalls to last two or three years."

Carol Bagby Chapman of Pfanstiel Road in "Sweet Owen" likes to stay in touch. "As Grandma fried little 'test patties' to taste to find out if the sausage was seasoned correctly, and was fixing supper, a truck pulled up, a man got out, and came to the door.

"He said, 'I'm from the Owen County Rural Electric Company, to hook up your electric like we promised.' He promptly got to work, while Grandma let out a whoop and set an extra plate on the table.

"Thank you, Mr. Rural Electric Man, wherever you are. And thank you, Rural Electric Company, for being there ever since."

OWEN ELECTRIC COOPERATIVE

Miles of Line:	4,413
Consumers billed:	56,256
Wholesale Power Supplier:	East Kentucky Power
Counties Served:	Boone, Campbell, Carroll, Gallatin, Grant, Kenton, Owen, Pendleton, and Scott

ADMINISTRATION

Chester Roland	1938 – 1944
Marvin Debell	1944 – 1953
Charles Richardson	1953 – 1967
Marvin Keith	1968 – 1978
Frank Downing	1979 – 2000
Robert Marshall	2000 – 2006
Robert Hood	2007 – Present

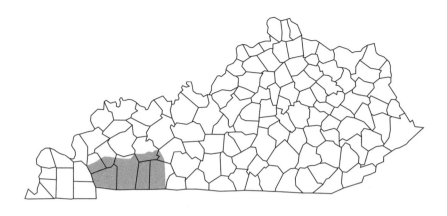

PENNYRILE
ELECTRIC

As we look to the future, technology will bring us even more opportunities to provide more dependable power at the lowest possible costs to our members, improving the quality of life for our employees and members.

Eston W. Glover Jr.
President and CEO
Pennyrile Electric

The historical marker on LaFayette Road at the Andy Haile farm tells the story: "1937 Pennyrile Rural Electric Cooperative Corporation—The dream of central-station electricity became a reality for 165 homes in the southern part of Christian County when a switch was thrown at this spot on the night of September 2, 1938. These 165 homes were the first members to be served by the Pennyrile Rural Electric Cooperative Corp. which had been organized, officially, on August 13, 1937.

"The infant cooperative was under the guidance of a Board of Directors composed of: W.E. Lacy, President; Thomas J. Lyne, Vice-President; G.W. Latham, Secretary-Treasurer; John L. Thurmond and

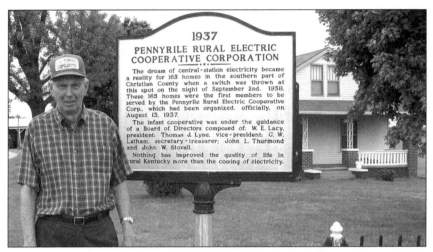

Douglas Haile, who still resides in the house where the first switch was thrown on August 13, 1937, stands beside the historical marker commemorating the event that lighted the first 165 rural homes in Christian County. (PLP photo)

John W. Stovall.

"Nothing has improved the quality of life in rural Kentucky more than the coming of electricity."

Mary D. Ferguson has written in the *Kentucky New Era*. Her description of the historic day is used here with her permission.

"A switch was thrown at the Haile residence on September 2, 1938, lighting the first 165 rural homes in Christian County with power from a substation at Masonville.

"Exie Heflin recalled that day, saying anticipation was great as the crowd of hundreds waited for the power 'to be turned loose.'

"Mrs. Heflin recalled how for one brother, the day meant all the soft drink he could consume, while a sister looked forward to using an electric churn."

We keep reminding ourselves of what J.K. Smith said in the Foreword: "Start with people—it's a people program."

Evelyn M. Boone knows the people of Todd County. She writes a column for the *Kentucky New Era*, and granted permission for its use here:

"Bringing electricity to our area was not so easy or readily accepted. This was a dramatic change in rural living, probably more dramatic than those involved could imagine. Pretty soon, contacts and contracts were made all across the area from Logan to Trigg counties. Shortly thereafter, lines began to be strung over fields and to houses. Some were suspicious about the wonder that was to come and become a normal part of everyday life.

"The story was told about the family that finally had electric lights in their home after years of dim light and darkness. The switch was pulled to light the house and the family continuously left the light on, for fear that if they turned the switch off, power might never come back on.

"Family life, as it was known before electricity was around, would never be the same.

"Stop and think about your life today. Grain bin blower, well pump, water heater, air conditioning, television, telephone, weather tracking computer, etc. Think how much easier family chores became.

"Probably the person in the family that saw the most change in daily life was the woman of the house. No longer did water have to be toted from the creek, spring, cistern, or well. No longer did the water have to be heated over a fire. How wonderful that abundant water could surge through the pipes. Food could now be stored in an icebox instead of salted. Food could also be cooked on a new fancy electrified stove instead of under the hearth.

"Because I grew up in the town of Covington, we had the luxury of electricity early on, but I remember as a child visiting my grandmother in the country who did not. One of the most exciting things to do on Saturday mornings was to clean the globes for the kerosene lamps. Grandma stated early in our visit to 'Be good and come Sunday morning, you will get to polish the globes.' My goodness, what a treat!

"Today's farm home is a place so involved in electricity—blenders, toasters, mixers, and microwaves. Oh my! Now the housewife can enjoy a nice cool house, a wonderful kitchen full of appliances, and multiple means of entertainment, including a computer that will link you to the world, an iPod for any type of music you may enjoy, and TV so you can watch your favorite shows, anytime you want.

"Electricity saved the homemaker from standing in a miserably hot kitchen, cooking all day long for many days to be able to feed her family in a satisfactory manner for the dreadfully cold winter ahead. Some homemakers still do a lot of sewing for the family, most of it done on an electric sewing machine that does much more than sew simple stitches.

"Oh, yes, let's not forget Saturday night baths. Some of you remember gathering the water and heating it. Usually the entire family would get to bathe before you, and sometimes you had to use the same water. Ugh!"

It was well-nigh time for a change.

Highlights of Board meetings of the Pennyrile Electric Cooperative from its conception tell a story of men and women in the area who brought about that change with great determination.

March 18, 1937—"A meeting was held in Elkton which was attended by Mr. J.W. Pyles, representative of REA, Walter Latham, Todd County REA Chairman, who appointed a committee consisting of the following: Mike Groves, Joe Hadden, T.G. Harris, J.W. Coots, Charlie Gill, L.W. Wright, Dudley Harris, S.D. Heltsley and E.O. Bell. Mr. Pyles requested that this committee work up some lines and he would return to Elkton on March 29 to inspect their work."

March 29, 1937—"Mr. Pyles approved the work completed and instructed the committee to put it in its proper form and submit it to REA for consideration by April 1, 1937."

June 4, 1937—"A meeting was called of the Pennyrile Electric Cooperative Association, a temporary organization with committees from Christian, Todd and Logan Counties. A temporary chairman from each county was elected as follows: Todd County—G.W. Latham; Logan County—Gaston Coke; Christian County—John L. Thurmond. Mr. Thurmond was elected chairman of this group. Representatives from the counties consisted of the following: Charlie Gill, R.C. Murray, John Stovall, Thomas J. Lyne, and W.E. Lacy. Also attending this meeting was Professor W.H. Hunt and Stewart Braybent, County Agriculture Agent of Todd County, who acted as temporary secretary."

August 4, 1937—"Approval was made of the Planters Bank and Trust Company and the Chanaberry Engineering Firm. J.T. Warren was considered for manager of the Cooperative and was employed in that capacity."

August 13, 1937—"The meeting was attended by Thomas J. Lyne, John Thurmond, Attorney Standard, Mr. Vennes of REA, Mr. Latham, Mr. Lacy, Mr. Stovall, Mr. Coke, and Mr. Braybent. Articles of Incorporation was discussed by Mr. Standard. The name of the Cooperative was selected and it was agreed at that time that the Cooperative would serve Christian, Logan, Todd, and Simpson Counties. It was agreed that the Board would consist of five members. The first board consisted of John L. Thurmond, Gracey; W.E. Lacy, Hopkinsville; G.W. Latham, Trenton; Thomas J. Lyne, Olmstead; and John W. Stovall, Adairville."

September 6, 1937—"Space was rented at the Elkton Bank and Trust Company as an office."

September 23, 1937—"Members of the Cooperative met and

What is an engineering department without a staff? In the beginning at Pennyrile there were (from left): Cliff Wood, Dan Hanbery, H.B. Parrent Jr., Martha Vancleave, Leslie Ellis, Thelma Massie, Richard Haddock, Lola West, and Harold Lee. (Pennyrile Electric archives)

Men of vision and determination to see the task beyond the present. The original Board of Directors at Pennyrile Electric were (from left) G.W. Latham, John L. Thurmond, W.E. Lacy, John F. Stovall, and Thomas J. Lyne. (Pennyrile Electric archives)

approved the by-laws and at a board meeting on that date, officials were elected as follows: W.E. Lacy, President; Thomas J. Lyne, Vice-President; G.W. Latham, Secretary-Treasurer. An engineering contract was signed with Ray W. Chanaberry."

November 12, 1937—"T.R. Gill was employed as bookkeeper-stenographer."

December 11, 1937—"Mrs. Edna Walker Petri, employed as stenographer-bookkeeper at a salary of $75.00 per month."

December 14, 1937—"The first construction contract with Arft-Killoren Electrical Company of Appleton, Wisconsin, was signed."

February 12, 1938—"The Board authorized the purchase of meters for the A section. A committee was appointed to see if TVA power was available."

April 12, 1938—"The Board authorized the signing of a wholesale power contract with the Ky.-Tenn. Light and Power Company."

June 13, 1938—"Edna W. Petri resigned and Miss Wilma Claiborne was named replacement. The first installment loan form was placed with two members of the Cooperative."

July 12, 1938—"The Board asked for bids on the first Cooperative-owned truck. Mr. William H. Waller and Mr. E.H. Rueppel of the State Electrical Inspection Bureau attended the meeting to discuss house wiring inspections with the Board."

July 27, 1938—"Discussion was had by the Board of the right-of-way trouble existing generally in Logan County."

August 12, 1938—"Bruce Wood of Russellville was employed as lineman for the Cooperative at a salary of $115.00 per month. Work was started on signing new members for the B section."

September 12, 1938—"The Board discussed sub-offices in Hopkinsville and Russellville. Eugene Lindsey was employed as lineman with a salary of $100.00 per month. The Board authorized the purchase of the second truck."

September 27, 1938—"The board discussed the possibility of purchasing the distribution system in Lewisburg. At the same

Linemen, right-of-way men, truck drivers, and ground men—all were needed to get Pennyrile going. In first two rows in no certain order were Bill Haddock, Mr. West, A.B. Cocke, Mr. Buie, Housewood Anderson, Cliff Wood, W.V. Anderson, Doug Hurt, Tom Cook, Cecil Hayes, and Hugh Foster. Back row L-R: Sam Wright, Andy Hester, and Tom Leavell. (Pennyrile Electric archives)

meeting approval was requested of REA for John O. Hardin as Attorney."

October 12, 1938—"John O. Hardin was approved as Attorney for the Cooperative. It was reported that the A allotment would be sufficient to build 280 miles of line rather than the 235 originally planned. Dorothy Sue Lawson was employed at a salary of $2.50 per day."

November 7, 1938—"Mr. Lacy and Mr. Hardin were to go to Washington to see about the purchase of Adairville, Lewisburg, and a small section of line near Edgoten on the B section. The Board authorized payment of $150.00 to the LaFayette Electric Company for the removal of existing lines in order that the Cooperative could serve the town."

December 6, 1938—"Employed Ray Chanaberry, engineer for the B section. Plans were made to work up a C section. The B contract was negotiated with Killoren Construction Company."

January 12, 1939—"Installation loan approved by REA in the

Time marched on and in the 1960s among those serving on the Board of Directors were E.G. Adams, E.W. Sweeney, R.C. Nichols, W.E. "Jack" Lacy, Robert K. "Bob" Broadbent, and John O. Hardin, Attorney. (Pennyrile Electric archives)

amount of $10,000 to finance members' wiring."

March 25, 1939—"Employed Mildred Johnson as stenographer-bookkeeper."

May 12, 1939—"Approved the purchase in the amount of $1,750.00 for 1 1/2 miles of line at Edgoten from the Cumberland Electric Membership Corporation."

May 26, 1939—"Executed B contract in the amount of $106,000.00 with the Killoren Construction Co."

July 12, 1939—"B change order in the amount of $163,000.00 to include Trigg County."

August 12, 1939—"Employed R.M. Tandy, right-of-way man at $85.00 per month. Also employed Wilson Wood at $85.00 per month. At the same meeting the engineering firm of Broaddus and Sullivan was employed and requested to prepare plans and specs for the C section."

October 17, 1939—"An offer of $3.00 was authorized by the Board to each member for the signing of a new member along existing lines."

November 16, 1939—"C allotment papers in the amount of $93,000.00 were executed."

November 25, 1939—"James T. Warren, Manager of the Cooperative, resigned and James O. Porter was employed as manager at a salary of $200.00 per month."

These were some of the building blocks leading to present time, 2008. President and CEO Eston Glover speaks for Pennyrile today:

"We started with 600 members in 1937 and today serve over 46,500 members in our nine-county service area, including a substantial portion of Fort Campbell.

"We switched wholesale power suppliers from Kentucky Tennessee Light and Power to the Tennessee Valley Authority in 1942.

"Our use of technology includes: automated meter reading equipment installed in nine months for 100% of our members; conversion of billing and accounting system to Southeastern Data Cooperative; conversion of paper maps to digital, and GPS laptop units in all vehicles.

"The future of PRECC is very good. Our load growth has been well above the norm in the past several years. We look forward to serving new facilities, industries, and residential neighborhoods, and also recognize tremendous opportunity at Fort Campbell. With new technology, we are now serving our members more efficiently than ever before. As we look to the future, technology will bring us even more opportunities to provide more dependable power at the lowest possible costs to our members, improving the quality of life for our employees and members."

PENNYRILE ELECTRIC

Miles of Line:	5,026
Consumers billed:	46,500
Wholesale Power Supplier:	TVA
Counties Served:	Butler, Caldwell, Christian, Logan, Lyon, Muhlenberg, Simpson, Todd, and Trigg

ADMINISTRATION

James T. Warren	1937 – 1939
James O. Porter	1939 – 1942
Howard Wiggins	1942 – 1977
John B. Mason	1977 – 1986
Quentis Fuqua	1986 – 2000
Eston Glover	2000 – present

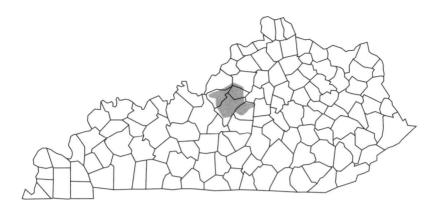

SALT RIVER
ELECTRIC COOPERATIVE CORPORATION

Salt River Electric has been a vital part of this community for seventy years.
Yet as important as our cooperative is to the people,
I believe the people are even more important to us.
It has certainly been a long and successful partnership.

J. Larry Hicks
President and CEO
Salt River Electric
Cooperative Corporation

"Less than twenty months following the historic creation of the Rural Electrification Administration (REA), a group of civic leaders from Bullitt, Nelson, Spencer, and Washington counties formed the Salt River Rural Electric Cooperative Corporation. Five of these men—M.W. Seay, W.F. Thompson, G.S. Greear, Ack Harned, and J.A. Wathen—became the founders, and original Board of Directors for the Cooperative.

"The first official meeting of the co-op was April 9, 1937."

Randy Burba, Vice President of Customer Service and Marketing at Salt River Electric, prepared the co-op's fiftieth anniversary report in 1987.

"As the first president of Salt River RECC, M.W. Seay played a vital role in executing the Cooperative's first REA loan of $315,000 on May 26, 1937. A farmer himself, Seay recognized the importance of electricity in agriculture. He was a member of the Nelson County Farm Bureau and served on the U.S. Department of Agriculture's War Board.

"The Bloomfield native served as president of the Cooperative for eighteen years until poor health prompted his retirement in 1955. His vacancy was filled by his son, David Seay, who served as a member of Salt River's Board of Directors until 1970. At his death in 1957, M.W. Seay left behind a rich legacy as the 'Founding Father' of the Cooperative.

"W.F. Thompson, who served as vice president of the Salt River RECC Board of Directors from 1937 until his death in 1951, was a resident of Spencer County.

"Active in both local and state farm-related organizations, he was a past vice president of the Kentucky Farm Bureau Federation, Director of the Southern States Cooperative, and a Director with the Burley Tobacco Cooperative Association. He, like the other Cooperative founders, envisioned a better way of life for farm families in this area.

"Grover S. Greear presided as secretary-treasurer of the Cooperative for thirty-seven years. At his death in 1974, he was the last surviving member of the original Board of Directors. Two times during his stint on the Board, in 1938 and again in 1939, Greear even served as temporary manager.

"Many years later he was elected to represent Salt River as a member of East Kentucky Power's Board of Directors. Greear, who wore many hats, was a former president of the Southern States Cooperative, Commissioner of the North Nelson Water District, and helped organize the Nelson County Burley Tobacco Cooperative in Bloomfield.

"Ack Harned served as Bullitt County's representative on the Salt River Board of Directors from 1937 to 1958. While a Director, Harned witnessed the energizing of the Cooperative's first lines on Feb. 10, 1938, giving 826 families electric service. He was also there for the first Annual Meeting, held June 25, 1938.

"Washington County's original representative on the Salt River

Board of Directors was J.A. Wathen of Fredericktown. He filled the position for thirteen years before migrating out of the county and resigning his post.

"During his tenure Wathen participated in setting the Cooperative's first residential rate ($3 per month for the first 40 kilowatt hours), and he also helped turn Salt River RECC into the third largest rural electric cooperative in Kentucky in terms of miles of line. Wathen was a member and former director of the Washington County Farm Bureau and served on the USDA War Board during World War II.

"This Cooperative has had its share of strong leaders, as a total of five men have operated the Salt River helm at one time or another. Each man, each leader, was special in his own way. But perhaps no manager was more influential, more instrumental in the shaping of this Cooperative than James S. Broaddus.

In 1946, the original headquarters office provided service to some of the richest agricultural land in the Commonwealth of Kentucky. How much richer it was as a result of the coming of rural electrification! (Salt River Electric archives)

"Jimmy Broaddus, an Oklahoma native, moved to Kentucky with his family at age sixteen. A resident of Boyle County at the time, Broaddus graduated from the University of Kentucky, and, with a degree in engineering, soon began to pursue a career involving electrical engineering.

"After working for several years with various consulting firms, Broaddus was named manager of the Cumberland Valley Rural Electric Cooperative in Corbin, in 1942. Two years later, he and his wife, Mary, moved to Bardstown where Broaddus accepted the post as manager at Salt River RECC. It was a job he would hold for the next thirty-two years.

"Under the guidance and direction of Broaddus the Cooperative prospered and flourished, and soon became one of the most successful, and financially sound, rural electric co-ops in the state.

"Pole top rescue" means one lineman helping to bring a stricken colleague safely to ground where resuscitation can restore the breath of life.Takes skill, strong will, and practice. (KAEC archives)

"In October, 1945, Broaddus orchestrated the company's move into new headquarters at 111 West Brashear Ave. in Bardstown. The new office building, completed and ready for occupancy in May, 1947, was constructed at a cost of $44,985.

"Armed with new headquarters and improved construction capabilities, Broaddus immediately began to expand the Salt River service area. With almost three thousand electric consumers already tied to the Cooperative's existing lines, there was only one way to expand the system—add more equipment. Broaddus, and the Cooperative, did.

"Late in 1947, construction was started on two new substations, one at Taylorsville and the other at Bardstown. By 1949, the Cooperative, with 4,500 members using an average of 154 kilowatt hours each month, was the third largest (in miles of line) RECC in Kentucky.

"By December, 1953, Salt River had constructed an estimated 1,600 miles of line across large sections of Bullitt, Nelson, Spencer and Washington counties.

"Also in 1953, under the direction of Broaddus, the Cooperative published the first-ever issue of a special consumer-oriented newsletter. The monthly, four-page fact sheet was eventually titled the Co-Amp News.

"In the spring of 1957, Broaddus announced plans for a major reconstruction effort. Included in the proposal were new substations at

Tornadoes attack utility poles as if they were matchsticks. Every story is a new challenge for the rural electric cooperative; move fast, correct the outage, and restore the customer's confidence. (Salt River Electric archives)

both Brooks and Mt. Washington. In addition, the Cooperative decided to enlarge the substation at Shepherdsville, and also began rebuilding power lines to Spencer County. A REA loan, totaling more than $1.1 million was also obtained. The money would be used to construct one hundred and sixteen miles of distribution line, providing service to eight hundred and seventy-four new customers.

"Four years later, in April 1961, Broaddus sought, and received from the Cooperative's Board of Directors, permission to begin a $1.2 million expansion project. The Co-op, at that time, served about eight thousand members with more than one thousand seven hundred miles of line.

"By March, 1965, Salt River had a total of ten substations, supplying power to 9,778 along 1,849 miles of line. And at the Cooperative's Annual Meeting, held August 3-4 that year, more than 18,000 people attended.

"The Cooperative, for the first time ever in 1966, awarded a pair of $500 academic scholarships to two high school seniors. The money enabled the students to pursue a career at the University of Kentucky's Colleges of Agriculture or Engineering.

"In May, 1969, again under Broaddus' direction, the

Improved technology is a constant necessity in times of stormy weather. The co-op has to budget for trouble. Expect the unpredictable... but real people have to answer the call. (Salt River Electric archives)

Cooperative decided to station six linemen in Bullitt County, where they were to live and work on a permanent basis. The move eventually led to construction of a branch office between Mt. Washington and Shepherdsville.

"With the addition of a new substation, constructed at Clermont in June, 1972, the Cooperative raised its total of power stations to twelve. The substations, in conjunction with more than two thousand miles of line, enabled Salt River to provide service for more than 12,000 consumers.

"From the moment J.S. Broaddus took over as manager at Salt River RECC, the Cooperative knew nothing but progress. But finally, after thirty years, the tide shifted for the veteran leader.

"In February, 1975, Broaddus asked for, and was granted, a leave of absence due to his rapidly declining health. With George F. Seyle serving as acting manager, Broaddus resigned his post June 30 of that year.

"An era had ended at the Cooperative, and so, too, did Broaddus' life less than three years later. He died Mary 29, 1978."

Throughout the years, J.K. Smith's goal of understanding and serving the people lived on despite the passing of each leader of the time.

Evan Keeling of Spencer County recalled: "Soon after the new line came through we switched to electric milking machines....Electricity enabled one man to do the same work as four or five....My Anna used to wash clothes all day, and I mean from sun up to sun down, on an old washing board....But as soon as electricity became available I bought her a brand new washing machine...and even a clothes dryer. Anna sure was tickled pink."

Charles "Bo" Bean was thirteen years old in 1938 when electricity finally reached Cox's Creek in Nelson County, about twelve miles north of Bardstown. Bo's Old Kentucky Home had been lit for years by a battery-charged thirty-two volt generator. The washing machine had a non-electric motor on it, and the refrigerator ran off kerosene. Mrs. Bean canned in summer, and daily in and out of season, and she had no choice but to cook on a wood stove—and the wood had to be split to fit, of course, and that was "pretty difficult."

"When electricity came, how much did you use?"

"Just the minimum $2.00 bill. Couldn't afford any more."

Now at eighty-two years of age, still living on the farm that's been in the family for 180 years, "Bo" Bean remembers the hot summer days when there was no air conditioning. "It was pretty hot in there," he calls to mind, but it was pretty cold in the wintertime when he went out to help with the milking. "The cow's teats had to be washed, but it was so cold the water was frozen, and we cried all over the place."

The Bean family milked six or eight cows, sold cream separated by hand crank, took what was left of the milk to slop the hogs. Bo says, "When I die, if I see a dairy cow I'll know I've gone to Hell."

"Kill hogs?"

"Everybody killed hogs."

"Baths?"

"There was a tub in the kitchen, and we heated water in a kettle on the stove. In the summertime we went to the creek. There were four of us kids."

"Drinking water?"

"We had a spring with the water piped in. Probably had germs as big as you and I—we got immune to it—a wonder we didn't get typhoid."

"Ice water?"

"We took ice from the creek in winter and dug a hole to store it in. We covered it with sawdust beneath an A-frame. We had ice cream! It wasn't until I was between fifteen and twenty years old that I knew ice tea didn't have to have sawdust in it."

"Outhouse?"

"Had a one-holer for the six of us. Three boys, one girl, and our parents."

"I thought most outhouses had three holes."

"We were poor people."

Organized in 1937, Salt River Electric today serves all or parts of Anderson, Bullitt, Jefferson, LaRue, Marion, Mercer, Nelson, Shelby, Spencer, and Washington counties. The main office is located in

Bardstown at 111 West Brashear Avenue, with branch offices in Shepherdsville, Taylorsville, and Springfield.

The Salt River, with its tributaries of Rolling and Beech forks, flows for about 140 miles from south of Danville till it empties into the Ohio River at West Point. The 2,920 square miles drained by the Salt are rich in agriculture and rural traditions.

Salt River Electric Cooperative Corporation means many things to many user/owners, but none more basic than the electric refrigerator on display at annual co-op meetings in the 1950s. (Salt River Electric archives)

Thomas Thornton remembers the dark days of Clermont in Bullitt County, where his father was "farmer in summer, hunted and trapped in winter." Looking back fifty-one years, he says it was "pretty rough." The old wood stove was "almost unbearable in summer." There were three sisters—two brothers died in infancy.

"The spring was located 200 yards from the house, and we packed drinking water. When the spring went dry, we had to carry water from a half mile from home. It was no easy life. We were poor but didn't think anything about it. We took baths in washing tubs, set them outside in summer. Didn't live too close to anybody. We managed. We called them the good old days."

Electricity came on in '43 or '44, but the Thornton family didn't get it until the early '50s, after they'd moved to Bardstown Junction. "Co-ops were one of the best things that ever happened...a great thing."

Thomas Thornton was about fifteen years old when the lights first came on, a "big deal now taken for granted." With a "little tobacco, corn, hay, and livestock," the Thornton family "specialized in staying alive."

It was a story repeated across the Commonwealth. Relatively short distances up and down the roads, urban areas were electrified—rural areas had to "take cold taters and wait."

President and CEO J. Larry Hicks spends his time today holding fast to the belief that "as important as our cooperative is to the people...the people are even more important to us."

SALT RIVER
ELECTRIC COOPERATIVE CORPORATION

Miles of Line:	3,872
Consumers billed:	45,389
Wholesale Power Supplier:	East Kentucky Power
Counties Served:	Anderson, Bullitt, Jefferson, LaRue, Marion, Mercer, Nelson, Shelby, Spencer, and Washington

ADMINISTRATION

R.F. Frizzell	1937 – 1939
Hugh Bell	1939 – 1944
J.S. Broaddus	1944 – 1975
George Seyle	1976 – 1978
Ken Hazelwood	1978 – 1987
George Seyle	1987 – 1990
George Mangan	1990 – 1995
Larry Hicks	1995 – Present

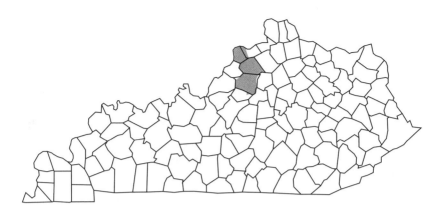

SHELBY ENERGY
COOPERATIVE, INC.

If you're consistent and tenacious in what you believe in,
it will work out. Sticking by what I believe in has been
why I am where I am today.

Debbie Martin
President and CEO
Shelby Energy Cooperative, Inc.

Ask Cora Mae Tipton, who lives on Washburn Road, about the day the lights went on in rural Shelby County, and she responds with two words: "Wonderful thing!"

She remembers going with her sister to the spring for water. "We went twice a day—she carried a gallon of water, and I carried half a gallon back to the house. We had [coal oil] lamps for light—it wasn't until after I married that I had electricity.

"We were the last people to drive horse and buggy—took three buggies to get us to church. We were Baptists, and we were raised to go to church every Sunday. We raised tobacco, corn, wheat, and hogs. Milked enough for the house, but Momma did sell a little cream. I was raised on milk instead of tea...we had milk every meal."

"How did you keep milk and butter cool?"

"We had a cellar."

In Mattie Shehan's family, there were eleven children—Dortha Mae, Leroy, Norris, Isaac, Randall, Oneida, Beulah, Ora and Nora (twin girls), Mattie, and Joe. They grew up in a tenant house without electricity in the Chaplin River corner of Nelson County, up in the Van Buren neighborhood (now served by Salt River Electric), a little place now covered over by the waters of Taylorsville Lake.

Mattie, born in 1931, remembers, "We walked to the Methodist Church about a mile away...didn't have a car, but we were raised to go to church every Sunday. We were tenants on a dairy farm. The man who owned it took us in. Neighbors always knew where to come for one-dollar-a-day help. I could milk ten cows a day myself. We carried water from a pond about one thousand feet away."

"Outhouse?"

"Two-holer. Rough in the winter time."

"'Sad' irons?"

"Had three or four on the stove. Mother ironed everything.

"On the day when we had electricity for the first time, Mother had started washing in the tub and on the board. When the lights finally came on, she left everything just the way it was, and she said: 'I'm using the electric washer today!'"

Wanda Ray was born on Vance Ridge not far from Port Royal, in the area where Wendell and Tanya Berry live. Wanda writes: "Living on various tenant farms in several counties, we had to make do with coal oil lamps as we called it. We carried water from the spring, well, cistern, a neighbor's, or the rain barrel."

A conversation with CEO and President Debbie Martin in her office at Shelby Energy, 620 Old Finchville Road in Shelbyville, confirms that gender has no glass ceiling in the cooperative rural electric structure. What counts is a passion for work, especially teamwork. (In August 1997, the name of Shelby Rural Electric Cooperative was changed to

Standing room only beneath the big cooperative tent. Who else would likely attract such a varied crowd? A community of Kentuckians electrified! (Shelby Energy archives)

Shelby Energy Cooperative.)

"How many jobs at Shelby Energy?"

"We have thirty-two. Most of the time when someone hires on here, they don't leave—we've had a few to leave but not too many."

"The major concern now is funding. It's the basic problem because how we're funded is going to directly impact what we charge our customers."

"What do you see as important for people to know?"

"How it will change. Fewer co-ops, mergers, consolidations. Our concentration is 6.5 customers per mile."

"Coal?"

"Such an abundant resource and with all the technical and scientific knowledge, someone out there can make this a clean source of energy...weigh protection of the environment and its cost to the people. More than seventy percent of our costs is paid to East Kentucky Power. You have to be very cost-effective with that thirty percent."

Debbie has been with Shelby Energy Cooperative for eighteen years. She's finishing her first year as President and CEO.

The underpinning of Shelby Energy, as in each of the Kentucky rural electric cooperatives, is the people, the consumer-owners, those who've lived their lives outside the urban areas.

George Busey—eighty-four, looking to be eighty-five—still farms on Harley-Thompson Road east of Shelbyville, two miles from Bagdad near where Guist Creek arises and flows west to form Guist Creek Lake.

He talked on the phone before going out on a zero-degree January morning to feed cattle. He shared a story about his mother when she insisted that they put electricity into the house.

"Father said, 'Why do you want this?' 'Because it's coming,' she said. She had the house wired in 1931-'32, wired in the middle of the Depression when the Bagdad School was built. They were redoing the house, re-plastering the walls, putting in hardwood floors, new windows. Mother insisted on wiring the house, father not so anxious. He didn't see it. But they roughed it in. Father asked a KU representative if an electric line could be run to the farm. Fellow said he would run the line for $2,500 dollars. Father said, 'You must be bidding on the farm.' KU didn't run the line. But Mother had premonition, the foresight to see this was the time for wiring the house.

"She had a stove with a boiler on the side, and when I was a kid and it was a cold morning I'd run around under the boiler—it was warm under there. In harvest time there'd be twenty to twenty-five in the house. She was a good mother—I was her only child. We did general farming, haven't changed format—cattle and tobacco."

"Did you milk cows?"

"We didn't have a dairy, but we took cream in a can to Bagdad and put it on the C&O [Chesapeake and Ohio] and shipped it to Cincinnati."

"Water supply?"

"Three wells. We drew water up with a rope. Had a cistern for roof water, had a chained pump."

"Was your outhouse a one-, two-, or three-holer?"

"Two-holer."

Joe Butler, eighty-six years old, lives at the top of a Trimble County hill overlooking the Ohio River. He's going in for another term on the

Shelby Energy Cooperative Board. He was fourteen or fifteen years old when electric power came to his farm home. His mother used a wood/coal stove. "It was a wonder how she did it. She also used a rocker washer and scrub board."

Joe Butler and his family lived in a log house with a cistern and a "nice spring house." They raised tobacco, cattle, wheat, corn, and hay, and had a small dairy. In winter, "It was bad milking six cows. Your hands were cold, and the cows didn't like it. But your hands got warm from milking cows. We sold cream in Milton. We had an old Delco battery, but it went bad and we couldn't afford to fix it."

"Why do you think the private power company didn't run a line to your farm?"

Being a contestant in the beauty pageant for Shelby RECC was exciting and caused all sorts of butterflies to come to life, but laughter was always the best medicine to dispel the nervous flutterings before the big announcement was made. (Shelby Energy archives)

"I couldn't understand it. Only two miles up to the hill."

"Outhouse?"

"Barnyard one-holer."

The history of Shelby RECC was often front-page news in the no longer existing *Shelby News*. In the April 2, 1959, issue: "The first pole in the five-county Shelby Rural Electric Co-op system was set up on the George Demaree farm, on the Burks Branch Road, in January 1938.

"The Shelby County line was the pilot project in a $200,000 grant from the federal government. It covered 213 miles and served 650 customers.

"Shelby's RECC prepares and mails over 4,300 electric statements each month. The billing is handled by Mrs. Louise Weakley and Mrs. LaVerne Murphy, who reads all meter cards sent in by members and prepares the statements.

"First milestone in the Shelby Rural Electric Co-op's system was reached May 20, when fifty-six miles of line in the northwest corner of the county received the first current.

"On August 19, 1937, the Shelby RECC board authorized the renting of office space, and a former one room store at 417 Main Street was rented from Mrs. S.W. Ford. Bill Dale hired Lou Russell Carr (now Mrs. James Lapsley) as bookkeeper and general office clerk to help him run the office and Shelby RECC was in business. A contract was awarded for the construction of 100 miles of line, and work was begun.

From mule power in the 1930s to horse-power in the mid-1980s, pole setting had come a long way, but without strong-hearted men like David Martin, Kenny Bennett, and Rick Shaw to get them set, the power would have been missing. (Shelby Energy archives)

"When the lines were energized in May, 1938, a lineman and a helper were added to the Shelby RECC staff. At first co-op billing in July, 1938, there were 108 co-op consumers using an average 40 kilowatt-hours each.

"In 1939 and 1940, residents of Henry and Trimble Counties, recognizing the value of the Shelby RECC program applied for participation in this electric service and construction of lines began in these counties. Residents of Carroll County soon followed the example of their Henry and Trimble County neighbors and joined the Shelby RECC.

"With the continuing expansion of Shelby RECC services and the accompanying increase in personnel, the RECC had outgrown its office facilities and the old Coca-Cola plant building at 2nd and Clay Streets was purchased from Jack Lawson in July, 1941. After the building was remodeled into four offices and a warehouse, the Shelby RECC staff moved in.

"By 1942, the Shelby Rural Electric Cooperative Corporation had a total of ten employees. The co-op now had 525 miles of line and was billing 1,185 co-op consumers.

"During the years of World War II, there was very little chance for expansion by Shelby RECC due to material shortages and government restrictions. However, in 1945, government restrictions began to loosen up. Shelby RECC by this time had a backlog of 1,500 applications for

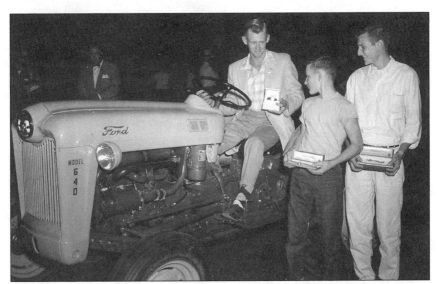

Sometimes the annual co-op meeting door prizes were worn on the wrist or they had to be driven away ... to plow, to mow, or to see how many more hours of daylight there will be before the lights go on. (Shelby Energy archives)

service that had accumulated during the war years.

"In 1946, to further facilitate its service, the co-op added a two-way radio system between the office and its service trucks.

"As Shelby RECC continued to grow and expand its services, it began to overflow its facilities. In the summer of 1949, three more offices, a garage and a warehouse were added to the building. The part of the building formerly used as a warehouse was converted to office space. There was now a staff of twenty-seven employees to handle billing and other co-op operations. The co-op had 1,000 miles of line and 3,600 consumers were receiving its services.

"In August 1950, William C. Dale, the first East Kentucky Power Board Chairman and the first Shelby RECC manager, one of the pioneers in the Kentucky cooperative movement, suffered a heart attack while waiting to speak at the annual meeting of the Shelby RECC and died shortly afterward. T.C. Long, Jr., who had been Shelby RECC office manager since 1946, was appointed acting manager by the board and, in January, 1951, he was officially named co-op manager.

"In 1956, the co-op again had outgrown its facilities. J. Quintin Biagi, Shelbyville architect, was employed by the co-op to explore the feasibility of further remodeling of the old Coca-Cola plant building. He advised against remodeling and suggested a new building be erected to accommodate the co-op's expansion. The co-op

In the latter 1930s, many people opened the door of the Shelby RECC office and walked right in. "You are a user? Well, that makes you an owner, which makes you a decision maker!" There's pride in ownership of a well-run organization.
(Shelby Energy archives)

manager and board of directors approved this plan. The building was to be financed partly by REA loan funds and partly by the cooperative's general funds.

"A six-acre site was purchased from Mr. and Mrs. E.P. Moynahan at Finchville Road and U.S. 60. Ground breaking ceremonies were held in May, 1958. Construction on the $240,000 building was completed in December, 1958. Shelby RECC personnel moved into the new building, which is one of the most modern, up-to-date office buildings of its type in the United States."

An unidentified newspaper clipping found in the archives of Shelby Energy Cooperative reported, "Contrary to a popular misconception, rural electric co-ops are not government agencies. Co-op members, who own the co-op, set their own policies and rates and operate the co-op just like any other business.

"Due to the increased use of electricity, the average cost to the consumer per kilowatt-hour has dropped from 6 cents in 1938 to 2.8 cents in 1958."

Dudley Scearce, now in his 80s, grew up on three-mile Clore-Jackson Road in northern Shelby County between Fox Run and Bullskin Creek. It was 1938 when the co-op brought electricity to the Scearce home. By then, Dudley was a student at Georgetown College. He graduated with a degree in economics and history but he went home during the summer. "From coal oil lamp to electric was a big improvement: refrigerator, toaster, but not all at once."

Dudley remembers his mother used two or three "sad" irons, and there was a galvanized tub for baths—had to use pressure for the hand pump—pump up and down to build pressure in a boiler. The cistern was the source of water.

The outhouse was a two-holer with magazines for reading and such—much crumpling, especially when the only paper was the *Saturday Evening Post* or the Sears Roebuck catalogue.

Mrs. Rosemary Campbell writes from Campbellsburg: "A few years ago, I drove my mother to her doctor. While sitting in the waiting room,

I overheard an older man talking. Seems he and the 'Old Lady' had been moved out of their mountain home to the city. He mentioned many things he missed and went on to say things he did not like.

"'Old Lady's' biscuits never browned good like in the wood cook stove.' The worst complaint he kept telling was that his white beans were not good. He said, 'She can cook all day on the beans, but I can taste that "leck-rick" in every bite I eat.'"

Maybe so, but five'll get you ten that if Old Lady were asked if she'd go back to the way things used to be before Shelby Energy came through, she'd give up cooking beans and her husband would have to find something else to chew on.

SHELBY ENERGY COOPERATIVE, INC.

Miles of Line:	2,050
Consumers billed:	14,954
Wholesale Power Supplier:	East Kentucky Power
Counties Served:	Anderson, Carroll, Franklin, Henry, Jefferson, Oldham, Owen, Shelby, Spencer, and Trimble

ADMINISTRATION

William C. Dale	1937 – 1950
T.C. Long Jr.	1950 – 1962
Thomas Barker Jr.	1962 – 1994
Dudley Bottom Jr.	1995 – 2006
Debbie Martin	2007 – present

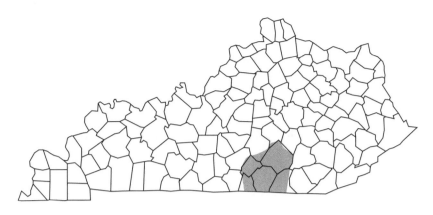

SOUTH KENTUCKY
RURAL ELECTRIC COOPERATIVE
CORPORATION

I think a major goal...should be to show to the best of our ability
what bringing electricity to rural Kentucky meant to those people
and to us still today.... The youth of today have never lived without the
convenience of electricity, unless there's an outage.

Allen Anderson
President and CEO
South Kentucky Rural Electric
Cooperative Corporation

Ninety-five-year-old Nadene Bertram of Windy near Otter Creek in
Wayne County has faithfully kept a personal journal. Part of her story,
written by Will Lindner in the June 2007 issue of *Kentucky Living*, is a
testament to the gradual coming of light in the Lake Cumberland area—
remote places from Alpha to Zula:

"A small group of farmers and businessmen had organized an electric
cooperative in 1938, which they called South Kentucky Electric
Cooperative Corporation. They obtained funding from the REA ...in
Washington, and, with considerable legwork and jawboning, recruited

239

sometimes-doubtful Kentuckians to become members and commit to a minimum $2.50 electric payment each month."

Today, $2.50 might buy three small candy bars, but not one gallon of regular grade gasoline. In 1938, $2.50 was big money. The Bertram family—like so many Kentuckians from Woodman on Lower Elk Creek in Pike County to Steppe Road along Washpan Lake in Fulton County—made do with the precious little they had. They waited patiently for that which should have been theirs. They may have taken a measure of pride in being poor, but this did not stop them from praying, "Let there be light."

Ewing and Nadene Bertram's first priority in the new age of electricity was a refrigerator. Before they bought the Philco they'd preserved their food as best they could, "the same way country folks had done it for years. We had a well," Mrs. Bertram says, "and we dropped our milk and butter down that."

In May 1967, the essential office staff consisted of the following who have been identified: (L-R front row) Sue Neikirk, Beverly Jasper, Hazel Brown, Joyce Alexander, Sam Hord, Manager, Carlene Jones, Iris Brown; (back row) Wanda Gilbert, Neta Slaughter, Cheryl Fisher, Jan Rainwater, Carolyn Wilson, Carolyn Hughes, Mabel Phillips, Chris Simpson, Cathy Miller, Frances Smith, and Linda Meece. (SKRECC archives)

Other purchases recorded in Nadene's journal:

Electric sewing machine, January 28, 1953, $114.85
Electric stove, September 13, 1954, $339.50
'52 Chevy, July 1955
Deepfreeze, 1959
New washing machine, 1965
Well pump, 1966
Color TV, 1970

Clarice Huffaker of Wayne County remembers: "Electricity came to the farm where I live in the early 1940s, when I was a child. It was a very exciting time. My dad, Earl Smith, became the first farmer in Wayne County to have an electric light with a timer installed in the chicken house, set to come on at 4:00 A.M., so the hens would awake and eat and produce more eggs."

Founded in 1938, South Kentucky RECC stalwarts who led the way to the fulfillment of letting there be light included: W. Dalton, Foxie Dunagan, L.G. Dunnington, R.H. Henninger, J.D. Mckechnie, Ray Oatts, G.P. Rice, T.M. Shearer, A.V. Stearns, and W.I. West. They and their families added up to many "sad" irons and outhouses in hot summer days and cold winter nights. That would have to change. And it did. Lights would come on and there'd be no turning back.

Today's South Kentucky RECC leader, President and CEO Allen Anderson, stands on the edge of progress honed by energies of the darkened past.

"One of the most significant achievements of SKRECC was the passage of a Mission Statement on October 3, 1999, and a Vision Statement on March 13, 2003. The mission statement says, 'South Kentucky RECC was formed for people, not profit.' Its mission is to 'reliably provide electricity and related services to its members at a competitive price, and to improve the quality of life in their communities.' The Vision Statement says, 'South Kentucky Rural Electric Cooperative Corporation's goal is to become a cornerstone of all our communities; to be recognized as an industry leader in service, professionalism, and competitiveness; and to be the energy provider of choice within our service boundaries.' The significance of both statements is they set the course for the cooperative to follow. SKRECC, for a time, had been in the cable television business, security business, etc., but with the implementation of these statements our goals became clearer and easier to follow. SKRECC strives to concentrate on providing electricity and to improve the quality of life in our communities through economic development, community participation, etc. We want SKRECC to be recognized as a cornerstone in each of our communities.

"South Kentucky RECC made a commitment in 2001 to help further the economic development of our communities. South Kentucky's leadership, a few years ago, realized they had reached a pivotal point in the co-op's history. One major way that SKRECC has tried to help improve the quality of life in its members' communities is through the establishment of an Office of Community and Economic Development. Established in 2002, this office has worked to find ways to partner with others on community projects and find available financial resources to help make these projects successful, from walking trails in community parks, business expansion and job creation, a senior citizens center, and new sidewalks, landscape, and underground utilities to downtown revitalization projects.

"SKRECC's Office of Community and Economic Development has been busy on a wide variety of projects. All these projects in some way improve the communities by creating jobs, expanding business, creating more inviting and appealing communities that others want to live in and raise their families. In addition, we feel this foresight will give our youth an opportunity to make the decision to stay at home and bring their knowledge back to our local communities, while becoming members of SKRECC.

"Since 2001, South Kentucky RECC has brought in $48,362,006.27 in funding for projects in its service territory, amounting to more than one thousand jobs in its area."

In 2005, for the first time in the co-op's history, the co-op

The substation construction crew is vital to the idea of teamwork. In the early 1950s in South Kentucky they were: (L-R front row) Jack Crawford, Eugene Cundiff, Clarence Dick, Willard (Blackie) Tarter, Herman Brawner; (back row) Chester Cain, Ira Baker, and Shelburn "Red" Adam. (SKRECC archives)

experienced one full year or 500,000 hours without a lost-time accident. Allen Anderson says this was a great accomplishment for the co-op. "We at South Kentucky RECC refocused our safety efforts and our safety program in mid-year 2000. We realized that we have jobs that can be very hazardous if we are not very focused and properly trained. We always tried to work as safely as possible, but we realized we had to do more. The first thing we did was to reaffirm the support of our management team and board of directors emphasizing the critical importance of safety and agreeing to support the necessary policies and financial commitment needed to make safety our #1 priority. This was an easy sell to our management team and board. We then adopted more stringent safety policies that included disciplinary actions to enhance our program. Our main goal, after all, is to see that all of our employees go home at the end of the day to their families safely and uninjured."

SKRECC was awarded the Kentucky Governor's Safety and Health

Award for 500,000 hours without a lost-time accident, but more important, employees are much better trained and are working safer.

"Another significant achievement at South Kentucky RECC has been the implementation of new technology," says Anderson. "In 1994, SKRECC implemented Supervisory Control and Data Acquisition or SCADA, which is used to monitor and collect data from substation control equipment. It has helped in two ways: we have access to real-time information that helps us to make important decisions more quickly at critical times (an example of this is having the ability to know quickly and precisely which sections of line at the substation feeder level are out of service so that we can allocate resources and get power restored); and it allows us to operate some special

In 1940, two men climbed a pole and installed wire and a transformer to bring electricity to the Acorn area. The inscription on the back of the photo reads, "Light was turned on Acorn P.O. and Store Jan. 27th, 1940. Taken by J. M. Mayfield." (SKRECC archives)

equipment without having to send people to the substations to do it. In recent years, SKRECC has put in place a new Outage Management System...In addition to the OMS, another new technology that we have instituted is GIS Mapping. For many years, our mapping system consisted of a physical map of our service territory on the wall of our dispatch center. With the new Mapping System, we can pinpoint the location of every pole, transformer, and member on our system. This also aids during outage situations because we can dispatch to that area quicker and more accurately. Along with the OMS and GIS Mapping, we have installed laptop computers in many of our service vehicles. That way, all of the information located in these two programs can be obtained instantly by servicemen while out in the field, saving time on the radio or telephone obtaining this information. We have also updated our radio system to a trunked radio system. This system not only allows much better communication throughout our entire thirteen-county service territory, but it connects all of our district offices by microwave radio. Eventually, SKRECC will have all computers, surveillance video, and telephone systems interconnected through this system. And the new radio system is convenient in that it can be taken from a vehicle and carried or used as a walkie-talkie at a job site.

"A historic step was taken in November 2007 when the city of Monticello's Electric Plant Board became a new part of South Kentucky Rural Electric Cooperative. By a two-to-one vote, around 3,500 more people became owners of their electric cooperative. With ownership has

And there they are — the thousands of user/owners of South Kentucky RECC at their historically immense annual meeting. Every vote counts in the cooperative concept and South Kentucky RECC boasts almost 67,000 members. (SKRECC archives)

"Let there be light" includes a Board of Directors willing to go the extra miles of service. In the mid-1970s, the SKRECC Directors were (L-R seated) Hugh Morrison, Bill Shearer, Rick Stephens; (standing) F.A. Smith, Grant Rice, Kenneth Hogue, and Jim Wilson. Not shown are Herman Schoolcraft, Manager, and Merril Harris. (SKRECC archives)

come opportunity to become involved in the decision making process, hallmark of the cooperative concept.

"We know that this will greatly change our future by adding more than 3,500 members to our lines, and tying the two systems together will provide more flexibility, more reliability, and more efficiency. The combination of the two systems will help us to realize an improved density per mile of line, an improved load factor, and allows us to experience more growth opportunity.

"We feel the future of South Kentucky RECC looks bright and will be one of positive growth, while keeping our members at the forefront of all that we do and all the decisions we make."

The decades following Nadene Bertram's earliest journal entries are filled with many new and beneficial developments at South Kentucky RECC. From Alpha to Zula, there's room for better lighted ways. It's well to remember that once there were no lights in rural Kentucky, and it would have remained that way were it not for the founders of the electric cooperative movement, strong people with a vision of the future.

It is well always to remember: South Kentucky RECC and the other twenty-five rural cooperatives have a heritage of lights in common: It has grown by and for the people.

Some waited longer than others.

Onza Wilson, for instance, didn't get electricity until 1951. It cost $50 to wire the house. The first purchases included an electric stove and refrigerator.

Maggie Russell remembers, "One bulb in the middle of the room thrilled me to death. Mother had irons on the stove, four irons and all those crinoline petticoats, you know, all trimmed with ruffles and lace."

Mrs. Wilma D. Hatter of Kings Mountain in Lincoln County: "I remember very clearly the night electric came on at our house. We were the second or third in our community, Walltown, with electric lights in our house. We had a small two-room house....Our first two electric appliances were a combination record player and radio and a refrigerator. Our family and friends came to listen to records and enjoy ice cream and ice cold water.

"Today we have electric in our home, barn, and garage. We have a range, refrigerator, washer, iron, T.V., radio, and small appliances. My husband has several electric tools in his garage. The old flat irons will be used as bookends. The washboard and ice box are gone. Our oil lamp sits on the table for emergency use only."

Well, Wilma's daydream for an electrically heated dream home came true: "I am eighty-one years old. Gave my brick home to my son and wife. Now I have an electric heated home with a whirlpool bath, handicap shower, and everything for handicap people...We had an ice storm one year...our electric was off two days. Our neighbor's electric was off over a week. I cooked breakfast for them—a widow lady and her crippled son.

"Thank God for our electric."

SOUTH KENTUCKY
RURAL ELECTRIC COOPERATIVE CORPORATION

Miles of Line:	6,600
Consumers billed:	65,900
Wholesale Power Supplier:	East Kentucky Power
Counties Served:	Adair, Casey, Clinton, Cumberland, Laurel, Lincoln, McCreary, Pulaski, Rockcastle, Russell, Wayne in Kentucky and Pickett and Scott in Tennessee.

ADMINISTRATION

F.D. Gregory(*)	1940 – 1942
Sam Hord	1942 – 1974
Herman Schoolcraft	1974 – 1984
Keith Sloan	1985 – 1999
Gary Cavitt	2000 – 2001
Allen Anderson	2001 – present

(*) Gregory granted a leave of absence to go to war. Sam Hord appointed superintendent in his absence. Shortly after returning to SKRECC, Gregory left the co-op.

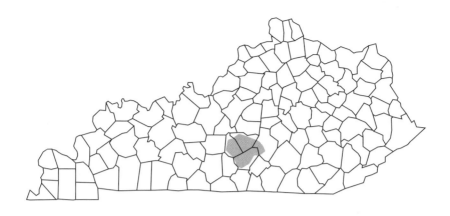

TAYLOR COUNTY
RURAL ELECTRIC COOPERATIVE
CORPORATION

We have an open door policy.
We have a real good bunch of employees.
We try to be good to our customers.
We try to do the right thing.

Barry L. Myers
Manager
Taylor County RECC

"We're a pretty down to earth system—slow to change—conservative. It's the type people who are here," says Manager Barry Myers. "I was born in Albany, but raised in Columbia. People are still down to earth, still got that unity. Mother used to say 'Watch out for the big me, little you—never get above your raisin'.'"

John H. Ewing Jr. was the guest columnist for the Golden Anniversary issue of *Co-op Hi-Lights*, April 1989:

"Look [back] and you will see that the nights across the United States were very, very dark in rural America fifty-three years ago. The private electric power companies refused to extend electric power lines to the

Taylor County RECC's first Manager, Earl Tomes,who served between 1938 and 1974, is surrounded by the first ladies of the office. They included (L-R seated): Ethel Parrott, Marjorie Hardwick; (standing) Lorena Ballard, Jane Ruth Fudge, Elizabeth Ingram, and Mary Jewell Vaughn Graves. (Taylor Co. RECC archives)

farm homes. Farmers with homes nearby large cities were asked enormous prices by the electric companies for extension of power lines. I know this because my family in Jefferson County lived one-fourth of a mile from the power lines, and were asked $5,000 to secure electricity. The power companies thought that farm families would not use enough electricity to make it profitable.

"Albert B. 'Happy' Chandler was Governor of Kentucky at that time and it was not long until Kentucky became one of the states to seek funds to establish rural electric cooperatives. C.V. Bryan, who was a dynamic county agent in Taylor County, and attorney Fred Faulkner of Campbellsville, were friends of 'Happy' Chandler, so they got on the bandwagon to seek a loan from the Rural Electrification Administration to establish the Taylor County RECC. They, in turn, requested county agents and civic leaders of Green, Adair and Casey Counties to work with them and others in Taylor to join in this very hard, slow and difficult job of organizing and establishing what is now known as the Taylor County Rural Electric Cooperative Corporation. H.H. Dickerson,

County Agent of Green County, organized the following Green County REA Committee: K.T. McMahan, F.T. Cantrell, Hugh Howell, Noah Blakeman, E.B. Judd, Austin Thompson, J.W. Moore, Eugene Shuffett, J.W. Pickett, Earl Henderson, Garnett Milby, W.A. Boyd, S.R. Reeves, Brady Milby, Maywood Mitchell, Bruce Clark, and Hugh Squires.

"These and many others in Taylor, Green, Adair and Casey Counties volunteered their services and went to work. The big jobs were to seek membership and also seek free rights of ways across the farms where the electricity was wanted. In order to secure a line, the requirement was five customers for each mile who would pay three dollars per month for a minimum of 40 watts per month. It was going to cost $500 per mile to build the line.

"The first Taylor RECC office was located in the Atlas Building which is now a part of the Campbellsville [First United] Methodist Church.

"The Taylor County RECC was officially organized May 10, 1938. The officers were: S.V. Kessler, President; K.T. McMahan, Secretary; and Directors F.L. Parrott, G.W. Penn, Paul Holt, W.J. Cundiff and R.L. Rogers. Mrs. Frances Sublett was employed as secretary, Mr. M. Lee Robinson was employed as bookkeeper—both in September, 1938. F.D. Gregory was employed as Project Superintendent in October, 1938, and served until Earl Tomes was hired as manager, January, 1940.

"It was January 1, 1939, when K.T. McMahan and I, on a cold, dark winter night, went to a meeting at the new RECC Office. The furniture was unique—nail kegs, orange crate boxes, a few old chairs, etc.

"The Taylor County RECC was well underway January 1, 1939, and everyone was anxious and looking forward to the time when the 'Big Switch' would be closed and electricity for the first time would light a few farm homes in Taylor, Green, Adair and Casey Counties.

"This was a self-help project and many farmers helped build the lines. Mr. Charles 'Nell' Netherland was using dynamite to blow out rocks in a hole for the line poles. It failed to explode—when Mr. Netherland peeked in, it exploded, he lost his eyesight! He was blinded; however, he ran a store along the Campbellsville (Greensburg) road for years.

251

"The 'Project A' Line in Green County came from Campbellsville down U.S. 68, then across my farm to Greensburg out to the Milton Vaughn farm and J.W. Moore. It crossed over to Ky. 61 by way of the Ellis Workman farm and went to Earl Henderson's farm at Summersville. The lines went out to S.V. Kessler's, Paul Holt's and other directions from Campbellsville and out Ky. 55 into Adair County.

"The big day arrived—February 27, 1939—when 57.4 miles with 116 customers were energized! This was a great occasion—Electricity had come to this area! Everyone was excited; and others immediately wanted electricity.

"The Taylor County RECC Office was moved to the corner of Main Street and South Central Avenue in a building formerly occupied by the Post Office, then to the building now occupied by Floy's and later to the present location in June, 1949. The office was destroyed by fire on November 20, 1952. The present building was built and moved into in May, 1954.

"S.V. Kessler had the honor of having the No. 1 membership. Ward B. Sleet was hired February 6, 1939, as the first lineman and Ewing Hoskins hired on January 2, 1939, as first maintenance man.

"The Co-op moved along very good and by May 15, 1939, there were three hundred and fifty customers.

"Earl Tomes was very aggressive in expanding the Co-op. He worked very cooperatively with me in helping the farm families secure electricity in Green County. We were very successful when a sincere farm leader came to my office and asked how he could secure electricity. I would hand him a batch of membership blanks and say all you have to do is secure five memberships for each mile from the nearest light pole to your home and also the right of way. L.E. 'Pete' Ronald did the good work out KY Highway 88; Marshall Shuffett secured the lines to Haskinsville; L.W. Shirley worked the roads to Mell for electricity; Ollie Larimore did the good work to get electricity to Bluff Boom. Asa Akin, G.W. Baxter and I worked the five miles from Allendale to Creal. R.B. Hancock and I walked through the mud with four-buckle overshoes to secure electricity to the Bramlett area. I would call Earl many times, and ask for electricity to be extended in difficult areas. I remember getting it extended to Sand Lick Branch and way beyond the Terrill School house.

"Green County farmers who have served as directors of the Co-op are: K.T. McMahan, Edmond McMahan, Garnett Milby, Shreve Loy, Chris Clark, Thomas Murray, Clinton Kelly and Pete Kessler.

"From Adair County, the following persons have served as directors: W.J. Cundiff, James C. Shirley, Bob Wolford, William Page, Richard Cobb, Frank Dohoney, David Bridgewater, Noel Harvey, William Harris and William E. Janes Jr.

"Taylor County has been represented on the Board of Directors by the following: S.V. Kessler, F.L. Parrott, G.W. Penn, Paul Holt, R.L. Rogers, Ezra Cafee, Howard Smith, Carl Smith, Clyde Akridge, Bobby Rucker and Rollin Minor.

"Directors serving the Cooperative from Casey County have been: Tommie Rodgers, Otha L. Compton, Hershel Porter, Vernie Foster, James Foxx and William L. Wethington.

"The faithful and excellent employees that have the longest tenure are: Ethel Parrott, forty-four years; James L. Shofner, forty-two years; Earl Hash, forty-one

"If we knew you were comin', we'd'a baked three more cakes!" Cake-baking contests at annual meetings for Taylor County RECC were made more fun by using electric appliances not previously available in rural areas of Kentucky. (Taylor Co. RECC archives)

years; Lucille Brockman, forty years and Anna Jean Rattliff, thirty-five years. Elizabeth Ingram (thirty-nine years) and Frances Sublett (thirty-four years) were two of the early, long-time, faithful employees, now retired.

"The advent of homes receiving electricity in the area, one hundred or two hundred homes every month, meant that there was an excellent market for all types of electrical equipment. I very well remember Woodson Lewis loading up a truck load of washing machines, electric stoves and refrigerators and heading to the country stores at Crail, Hope, Mell, etc., and selling them as the people came to the store.

"The Taylor County Co-op now [in 1989] has 2,618.4 miles of lines [which] if stretched in a straight line would reach from Washington, D.C., to San Francisco. There are 16,813 customers, who use an average of 1,038 kilowatt hours, with an average monthly bill of $61.36."

"Lucille and Carl Lemmon married in 1945 and moved to Pellyton, Kentucky, and didn't have electricity. [Pellyton is in Adair County, close by Kentucky author Janice Holt Giles' *Little Better Than Plumb* cabin on

Once hired, employees are fiercely loyal to their co-ops. Many of those in this office on April 13, 1954, were there in the beginning: (L-R front row) Elizabeth Ingram, Lorena Ballard; (second row) Anna Jean Rattliff, Exie Burris; (third row) Ethel Parrott, Helen Bowles; (fourth row) Margaret Tomes, and Mabelle Bright.
(Taylor Co. RECC archives)

"Now, just where do you want those poles set?" "Those lines strung?" "Those connections you want made safe and secure?" These are the original "can-do" guys of Taylor County RECC. None better. (Taylor Co. RECC archives)

Spout Springs Branch.] Lucille and Carl used coal oil lamps, cleaning the globes every day and filling them with kerosene each day. They did their washing on a washboard and used 'sad' irons. They kept their milk cold by putting it in a jar inside a bucket and lowering it into the well. By 1946, their first child was born. Carl had to help do the washing on the washboard and drawing the water from the well.

"In 1947-1948, they got electricity and the first things they had with electricity were a few plugs, overhead lights, iron and wringer washer. Around 1950, they built a house, still cooking on a wood stove. It wasn't until about 1955 that Lucille and her mother-in-law decided to buy an electric stove. After a day of washing the clothes, a day was set aside for the ironing. The clothes were made from mostly cotton and they wrinkled. They used starch to make the clothes look better. Starch was made using flour and water and cooking it on the stove. She didn't use that too long, Faultless [brand] starch arrived on the market, but it still had to be cooked. The recipe was on the back of the box. You had the choice of light or heavy starch. Clothes were sprinkled using a shaker bottle, rolling the clothes into a roll and putting them inside a folded sheet to keep them damp. After sprinkling them, there was a time period they had to be ironed or they would sour or dry out." (Interview by Ann Beard, Manager Member Services, Taylor County RECC.)

"Gene and Ida Walker are both eighty-two years old; Ida was raised in Louisville and had electricity. She and her first husband moved to Green County, Kentucky, and lived in a small house with no electric power. It was around 1950 when her home was first connected. The first things she got that required electricity were an iron, wringer washer, and refrigerator. They milked ten cows by hand, kept the milk cool in washing tubs of cold water from the cistern. She said the milk still spoiled when temperatures got too hot. They sold the milk to a cheese company in Horse Cave. They had to keep the milk and butter for personal use in the spring. She said a lot of people kept the milk in a bucket in the cistern, but she was always afraid that she would spill the milk into the water and would ruin the drinking water.

"As a child in the mid-1930s, Ida remembers a neighbor getting electricity. A mile of line needed to be run to the neighbor's home. Ida's father agreed to dig the post holes for the electric lines for seventy-five cents per day. Her family didn't have electricity at that time because they couldn't afford it.

"They used coal oil lamps and a battery radio. The radio was turned on for news and as soon as the news was over the radio would be turned off, conserving the battery for the Grand Ole Opry. She had an old treadle sewing machine that was a chain-stitch machine. She still has it but it needs a belt. She used a simple wood stove at first, then a coal oil stove, and a bottled gas stove. She finally bought her first electric stove in the mid-1960s that had the deep well cooker. She still has the pan that went with the stove. Ida got running water in 1969." (Interview by Ann Beard)

"In 1953, Betty J. Williams' first year of teaching was in a one-room schoolhouse. Posey School was located near Clementsville, Kentucky, and had about twenty-five students ranging from first to eighth grades.

The school didn't have electricity, and Mrs. Williams remembers having pie suppers to raise money to have electricity installed." (Interview by Ann Beard)

Eloise Elmore remembers: It was March 15, 1939, when "electricity was turned on in our house in the Chestnut Grove community in Taylor County. Daddy said, 'The lights are so bright my eyes hurt.'

"On March 16, we got our electric radio hooked up. Before that we had a battery radio.

"April 27, we got our new washing machine, which saved Mother a lot of hard work. Mother had six children.

"We got an electric motor cream separator—before that we turned the wheel which made the dish separate cream from milk.

"In 2008, electricity helps widow women live more easily. No one has to carry water and wood."

Ed Hudgins of Matney Road in Taylor County recalls that "In the year of 1953, my grandfather was approached by Taylor County RECC

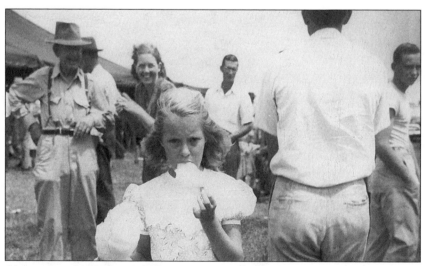

Mmmmmm. Cotton candy spun miraculously by an electric machine was a special treat for kids in the rural areas of Kentucky at early RECC annual meetings. What's a little sticky sugar on faces and fingers? (Taylor Co. RECC archives)

about running an electric line to his house from about 300 yards away from the end of the line at the Green-Hart County line.

"The two linemen and my grandfather, John Lonzie Hudgins, packed the pole from the gravel road in front of the house to the back and then the two linemen and my grandfather, using a post-hole digger with roughly twelve foot handles, proceeded to dig a deep hole for the pole. Two wires were pulled to the top, threaded through insulators and pulled tight.

"This was the manual way of providing electricity for my grandparents and rural America, with fond memories that make me feel like a kid again!"

TAYLOR COUNTY
RURAL ELECTRIC COOPERATIVE CORPORATION

Miles of Line:	3,136.28
Consumers billed:	24,679
Wholesale Power Supplier:	East Kentucky Power
Counties Served:	Adair, Casey, Cumberland, Green, Hart, Marion, Metcalfe, Russell, and Taylor

ADMINISTRATION

Earl Tomes	1938 – 1974
Barney Blevins	1974 – 1978
Bill Nixon	1978 – 1985
Barry L. Myers	1985 – present

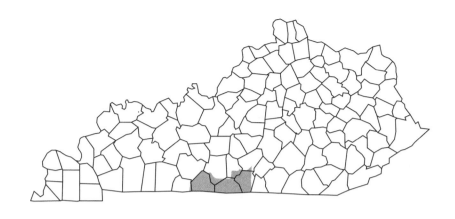

TRI-COUNTY ELECTRIC
MEMBERSHIP CORPORATION

*Tri-County Electric has experienced steady growth
throughout the past seventy years bringing current
membership to approximately 50,500.*

Paul Thompson
Executive Vice President and
General Manager
Tri-County Electric

In 1935 the Rural Electrification Administration was created by
President Franklin D. Roosevelt to bring electricity to rural America.
Wasting little time, Will Hall Sullivan, the founder of Tri-County
Electric, and the original Board of Directors—William Parker, C.A.
Hammond, Dr. J.Y. Freedman, and Lee Hanes—organized Tri-County
Electric in 1936 to serve rural areas in Macon, Trousdale, and Sumner
counties, Tennessee.

From his office at Tri-County Electric's headquarters in Lafayette,
Tennessee, Paul Thompson, Executive Vice President and General
Manager of the cooperative, highlights the history and accomplishments
of Tri-County Electric.

"On Christmas Eve, 1937, the first eighty-five miles of line were energized. In 1939, Tri-County Electric Membership Corporation purchased the holdings of the Tennessee Electric Power Company, which included lines in Celina, Tennessee. The 1940 purchase of the Cumberland Public Service Company expanded Tri-County Electric's service area to include Burkesville, Tompkinsville, and Edmonton, Kentucky. A portion of the Kentucky-Tennessee Light and Power Company was purchased by Tri-County Electric in 1942 allowing us to serve the Scottsville, Kentucky, area. Acquisition of the Scottsville service area brought Tri-County Electric's 1943 membership to 7,000.

"Tri-County Electric Membership Corporation has experienced consistent growth over the past seventy-one years. Our wholesale power is supplied by the Tennessee Valley Authority. We currently have seventeen substations and over 5,400 miles of distribution line, covering a service area of roughly 1,800 square miles. Tri-County Electric serves all or part of the following counties: Allen, Cumberland, Metcalfe, Monroe, Adair, Barren, Clinton, and Warren in Kentucky; Macon, Clay,

Will Hall Sullivan and one of the original Board members, Lee Hanes, on the front porch of the Cordill Hull farm home when it was energized. Secretary of State Hull was awarded the Nobel Peace Prize in 1945. (Tri-County Electric archives)

Trousdale, Sumner, Jackson, Overton, and Smith in Tennessee. We are the fifth largest cooperative in Kentucky or Tennessee and provide a wide range of services to our approximately 50,500 members-owners."

Although there are many areas Paul Thompson emphasizes in his list of achievements for the cooperative, he believes there are two that stand out: "In the fall of 2002, Tri-County Electric became the first cooperative in Tennessee or Kentucky to implement an automated reading system."

The second focus began as an effort to provide severe weather notification

Tri-County Electric's founder, Will Hall Sullivan, a visionary. Where others just dreamed, Sullivan planned his dreams and worked his plan for the betterment of those whose lives were brightened by his visions. (Tri-County Electric archives)

to the cooperative's rural, unserved area and resulted in Tri-County Electric being awarded the prestigious Mark Trail Award by the National Weather Service in 2004. "Tri-County Electric worked very hard to secure a NOAA Weather Radio Transmitter to provide our member-owners the technology that would warn them of the possibility of severe weather, Amber Alerts, and other alerts deemed important enough to be issued by the government," says Thompson.

According to the National Weather Service, the Mark Trail Award is presented to "individuals, state, county, and municipal governments, non-government organizations and corporations for noteworthy gifts, community action, and individual or group response to a NOAA Weather

Radio warning that exemplifies its lifesaving benefits." In addition to applying for and securing the grant for the weather transmitter, Tri-County Electric donated NOAA Public Alert radios with SAME technology to every school, superintendent of schools, hospital, and nursing home in its service area.

At the time, Tri-County Electric could not have known the value the weather transmitter would have on Tuesday night, February 5, 2008, when a tornado cut a forty-three mile path through the center of Tri-County Electric's service area and devastated portions of Sumner, Trousdale, Macon, Monroe and Allen counties.

In approximately forty-five minutes the storm destroyed more than 400 transmission/distribution poles, resulting in loss of power to 22,500 of Tri-County Electric's member-owners. By mid-morning the following day more than 100 additional linemen and support staff from neighboring cooperatives and contract companies were on the scene and assisting Tri-County Electric personnel in restoring service.

The cooperative spirit of neighbors helping neighbors was the key to rural electrification in America during the 1930s. That commitment and dedication to service is still alive today.

"You truly understand the value of being a part of the co-op family in a crisis situation," says Paul Thompson. "Everyone we contacted responded immediately and others called us on a daily basis to offer their support. Electric cooperatives talk about putting our member-owners first and, in our case, that's really easy to do because our Board of Directors and employees are local. The member-owners are our family and friends. They belong to the same churches and civic organizations, their children attend the same schools and play on the same sports teams. We had employees who suffered severe damage and others, total losses. Some of them worked around the clock with their fellow employees to repair, rebuild, and restore service to our member-owners who depend on electricity for their way of life. They demonstrated a commitment to service above self by looking beyond their own situation and focused on putting our member-owners' needs first.

"In 1936, Will Hall Sullivan's vision was to create a corporation that would operate on a cooperative basis, not for profit, and for the benefit of its members. Tri-County Electric maintains that same commitment today to improve the quality of life in the communities we serve in everything we do.

"Will Hall Sullivan was born in Lafayette, Tennessee, on January 25, 1911, to William J. Sullivan and Dona A. Sullivan. Will Hall dropped out in his third year of high school to assist with the family farm due to his father's failing health. In 1929, Will Hall developed a process to mechanize a horse-drawn planter. Will Hall's name was engraved on a bronze plaque in the Machinery Hall at the Chicago World's Fair recognizing him as one of the World's Pioneers of Horseless Farming.

"Will Hall saw firsthand the hardships of farm life and felt electrification of this rural area would be the best possible way to improve the quality of life for these farming communities. Will Hall's visits to Washington to request federal financing from Sen. George Norris is believed to have led to the act that created the Rural Electrification Administration. This earned Sullivan the distinction of being called 'the father of the Rural Electrification Administration,'

Membership meetings have been held in lots of places, but in the late 1930s, Tri-County Electric's founder and first General Manager, Will Hall Sullivan, addressed the members at one of the first annual membership meetings from within a boxing ring! Nowhere better to "duke out" any problems that might arise.
(Tri-County Electric archives)

which helped bring modern conveniences to many homes across Kentucky, Tennessee, and the nation. Will Hall Sullivan can no doubt go down in history as the man who has had major influence on the development of Macon County, middle Tennessee, and southern Kentucky."

In the January 11, 1988, edition of *The Macon County Times*, the paper printed "A Memoir by Carlos Smith." It was he who assisted Tri-County Electric's founder, Will Hall Sullivan, in the original mapping of the cooperative, and a portion of Mr. Smith's memoir is reprinted here with permission.

"Back in 1936, my wife Ethel operated a produce house and cream-buying station located in the old Johnson building on Jail Street here in Lafayette. We opened the first cream-buying station in Lafayette to make ends meet. I did a little sign painting in the back room. I would paint a sign for businesses around town for fifty cents or seventy-five cents. I would paint a sign on a spare tire cover, which most cars had on the rear.

"Will Hall Sullivan sold us cream, and he saw some of my work, so one day he asked me if I could draw some maps for him of roads and power lines. I said I would try.

"Will Hall arranged for office space over the Citizens Bank Building, which was where Buddy Jent's Drug Store was at that time, so I agreed to help out. I became very interested in his project, so the best I can remember, we worked three or four nights each week. We would meet up there after supper and work until nine or ten o'clock.

"The material we drew our maps on was cardboard. Will Hall would work in the field in daytime and bring his fieldwork in for me to put them on the map. When we had our map work finished, Will Hall took a trip to Washington with his big roll of cardboard maps."

Tri-County Electric "old timers" will not soon, if ever, forget those days of Sgt. Alvin York (1887-1964), the Tennessee World War I hero who grew up sixty miles to the east of Lafayette before there was anything as simple as a light bulb in a rural house, school, or church.

John Fitzwater, eighty-seven years old, was employed by Tri-County

These fourteen men—one of the early Tri-County Electric crews—are examples of some of the stalwart individuals that hacked their way through rights-of-way, slogged through mud to pull lines, brought power to many a home, and smiles to lots of faces. (Tri-County Electric archives)

in the summer of 1963, and he worked until the last day of 1984. "My father, John Roger Fitzwater, was a Methodist circuit rider. I was eight or nine years old—had a twin brother and two sisters. I rode with my father, rode on saddlebags—roads were impassable—had to travel behind him on the horse. I was born in 1919, traveled with him in the '30s. I was a meter man, transferred from Clark RECC in '63...was with Clark twelve years. There was a winter storm, and I set out with another man. [He was hit with] 7,000 volts, knocked unconscious, hanging by his belt, I had to resuscitate him and keep him alive until someone else came and helped get him down. I belong to the Turtle Club. Someone dropped a compression wrench on my hard hat...the hat saved my life. The club is not very big.

"The real meaning [of having electricity] is good roads. Electricity paved the way for better roads. Electricity is the backbone of this country."

Tom Gulley, seventy-seven years old, was employed by Tri-County in November of 1948, and he worked until the last day of 1993. "I had some close calls...helped set poles by hand. Worked all or parts of fifteen counties in Tennessee and Kentucky."

"How many in your family?"

"Twelve children—nine boys and three girls."

"Baths?"

"Lucky to get them."

Ollie Harper, ninety-two years old, worked for Tri-County for thirty-five years.

"Mighty hard times—could hardly get hold of a nickel. Father was the roughest, but he could take it. Thirteen children—I was the oldest."

For more than 70 years, digging in the heels, climbing the poles, tightening the safety belts, wearing thick gloves, knowing what to touch and what not to touch, the work still goes on and on, 365 days a year. (KAEC archives)

"Baths?"

"Rounded 'em up like sheep—seven boys and six girls. Had a curtain."

"Your mother?"

"Had her hands full—three to four 'sad' irons. Had a springhouse. We carried buckets of water up a big, steep hill."

"Still, not everybody wanted electricity?"

"There was a little hollerin' [about hooking up to electricity]. Father had brothers who didn't want it. We raised corn and tobacco. Had a garden. It took a whole lot to eat. Had plenty to eat but didn't have any money. Most I ever had were six dollars."

"Your worst story?"

"I was working with Indianapolis P&L for eighteen months. Came back on vacation—was making $1.07 an hour in Indianapolis and Will Hall Sullivan was offering $.50 an hour for construction work here. I had to go into the army anyway, but I came back and stayed here after the war.

"In '51 there was a bad ice break—in the ridge through here all the poles were snapped."

"Much of a problem with people stealing electricity?"

"Stealing electricity? You'd be surprised the different ways these rogues can steal electricity. Some turned meters upside down."

The "E.E. Thomas Papers, 1905-1934," in the Tennessee State Library and Archives, contain a description of an early one-room school in Tennessee. No mention of electricity. Wood stoves in winter, open windows on hot days, drinking water from a spring: a picture repeated throughout rural America. Not until the mid-1930s did life make a dramatic turn with the coming of electricity in rural areas where there'd been none before.

Those primitive days have come and gone. The present and the future are in the hands of young leaders with a new commitment to excellence—people connected from Will Hall Sullivan to Paul Thompson. "I believe our dedication to the communities we serve through economic development and scholarship programs and support of civic organizations, coupled with a continued commitment to follow the cooperative business model of putting our member-owners first in everything we do while maintaining our primary focus of delivering high-quality, reliable, and affordable service every day to every member-owner will continue to bring success to our area."

TRI-COUNTY ELECTRIC
MEMBERSHIP CORPORATION

Miles of Line:	5,438
Consumers billed:	50,438
Total KY Consumers billed:	24,615
Wholesale Power Supplier:	TVA
Kentucky Counties served:	Adair, Allen, Barren, Clinton, Cumberland, Metcalfe, Monroe, and Warren

ADMINISTRATION

Will Hall Sullivan	1937 – 1946
Raymond Forkner	1946 – 1947
Ewing Hoskins	1947 – 1965
Charles Mayhew	1965 – 1976
Ottis Jones	1976 – 1979
Charles Mayhew	1979
Paul T. Lee	1980 – 1983
Charles Mayhew	1983
Jack Dillard	1984 – 1991
Wayne Wilkins	1991 – 1992
Kelly Nuckols	1992 – 1995
Gerald Freehling	1996 – 1997
David Callis	1997 – 2001
Paul Thompson	2001 – Present

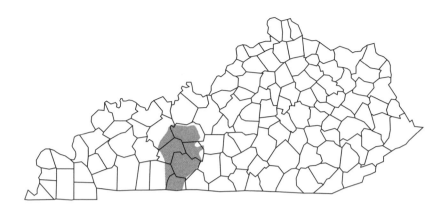

WARREN RURAL ELECTRIC
COOPERATIVE CORPORATION

*The intent of Warren RECC has always been to provide
the members with reliable power at a competitive price
and service when there is a problem.
That's what it's all about.*

Gary K. Dillard
President and CEO
Warren Rural Electric
Cooperative Corporation

"In the mid-1930s, only three of every one hundred farm homes in Kentucky had electricity—just slightly more than three percent—and these had paid dearly for this convenience."

A history of Warren RECC was prepared in 1971 by the late William O. (Bill) Skinner for *Rural Kentuckian* editor Gary Luhr's Fiftieth Anniversary project.

"There were several in this southern Kentucky area who wanted to take advantage of this 'new way of life' which was offered through REA and TVA. They were ready to contribute of their time and energy to promote interest in a rural electric cooperative.

"Among them were Warren County Extension Agent A.E. Ewing, Anthony J. Warden, one of the most ardent supporters of the program and who, as a charter member of the soon-to-be-formed Warren RECC board of directors, did not miss a single meeting until his death in 1957, and W.W. Chapman, also a charter board member.

"On April 8, 1938, Warren Rural Electric Cooperative Corporation was in business. Making up the initial $35 were members of the Cooperative's first board of directors—Lester Wright, Smiths Grove; Cecil Elmore, Rocky Hill; L.H. Hildreth, Smiths Grove; Anthony J. Warden, Route 2, Bowling Green; W.W. Chapman, Route 3, Bowling Green; Dewey Strickler, South Union; and O.H. Park, Rural Route, Franklin.

"On May 5, 1938, the Board of Directors called for a general meeting to be held on May 24 at 1019 State Street in Bowling Green...With an office in which to meet and two employees on the payroll, the way was opened for a final organizational meeting of stockholders [members] on October 1, 1938. A debt limit of $1,500,000 was set up, and the group authorized construction and operation of electric generating, transmission, distribution and service facilities at such places and along such routes as are needed.

"Four counties—Warren, Edmonson, Simpson and Logan—were set out as the area in which the Cooperative would operate initially.

The day toward which the cooperative had worked for many months and toward which the rural family had looked for many years was fast approaching. Energization of that first 130-mile line was near at hand. And for those who had worked and waited and weathered many disappointments, it could not come too soon.

"March 3, 1939, is a day that will long be remembered by many rural families in southern Kentucky, for it was on that date that electricity came to 325 homes along that first 130-mile network of lines.

"Celebrations were spontaneous. Many observed the occasion by tossing their kerosene lanterns out the door. For where there once was only darkness or, at best, a flickering flame obscured by a smoked chimney, there now was light—electric light.

"The convenience of electricity, which the city resident had known for many years, had now come to the country, at least in part. And the miles of line through which the electricity coursed were to grow and stretch out like a spider web in all directions so that even more of those who previously had been denied could enjoy the benefits of cooperative power.

"Virtually every home had been equipped with only a thirty-amp service, and for many, this was soon to be crowded to capacity as more home conveniences were added.

"Electric power was now available outside the city, but to all too few. As rapidly as work could be performed and as finances would allow, power lines were extended into out-of-the-way places. Construction moved forward outside the boundaries of Warren County to include sections of Edmonson and Butler Counties.

The first Warren RECC Board of Directors, at the first annual meeting in June 1939, in front of the co-op's first office building. Left to right they are: O.L. Parks; Cecil Elmore; a representative from REA; M.V. Hatcher; Lester Wright, first Manager; Jimmy Broaddus; W.W. Chapman; Anthony Warden; and Dewey Strickler. Absent when the photo was taken was L.H. Hildreth. (Warren RECC archives)

"On May 25, 1942, the Board approved a resolution authorizing the purchase of Kentucky-Tennessee facilities in Warren, Edmonson, Butler, and Barren Counties for a sum of $81,900. Included in this purchase were all consumers along Louisville Road north of Bowling Green to Park City, the towns of Park City in Barren County, Smiths Grove and Oakland in Warren County, and Rhoda, Rocky Hill and Brownsville in Edmonson County.

"On June 1, 1942, power generated by Tennessee Valley Authority came to a large portion of the Warren RECC service area. All of that area south of Green River was included in the switchover, while the remaining area was supplied by power purchased by the Cooperative from Kentucky Utilities Company.

"In August, 1942, the Cooperative moved its headquarters from the American Legion Building at 826 State Street in Bowling Green to newly leased facilities at 8th and State Streets.

"Authorization to increase the debt limit by $50,000 was voted by the Cooperative's membership at the Annual Meeting of June 26, 1945. This paved the way for a $270,000 loan from REA and backed up a membership resolution to construct and operate additional facilities in Barren, Butler, Edmonson, Grayson, Logan, Ohio, Simpson, and Warren counties.

"The city of Morgantown, county seat of Butler County, began negotiating with Warren RECC for service in February, 1946, and this municipally-owned system was purchased by the Cooperative on November 25, 1947, for $12,000.

"Highlight of the meeting [June 29, 1946], held at Snell Hall on the Western State College campus, was a vote to increase the Cooperative's debt limit to $4,000,000 to meet immediate expansion needs. Even with this financial block removed, however, it would still be close to two years before a badly needed postwar construction program could be started."

With the resignation of General Manager Lester Wright in 1948, "Mrs. Ollie Mae Roberts, who had been an employee of the Cooperative since March, 1939, and was destined to become Warren RECC's first 30-year employee, was designated acting manager.

Just look where these ten men in the '30s led Warren RECC! From a beginning of 130 miles of line for 325 homes to now over 5,500 miles of line and more than 58,500 homes. What a legacy! (Warren RECC archives)

"The search for a new manager who could lead the Cooperative through vital years ahead led to Charles M. Stewart, then manager of Farmers Rural Electric Cooperative at Glasgow, Kentucky. Mr. Stewart accepted the appointment and took over his new position on September 15, 1948.

"The year 1948 was to be an important one in the history of Warren RECC for reasons other than the employment of a new manager.

"The expansion of its service area and the demand for a closer-knit program brought about the establishment of a branch office at Franklin in January with Guy W. Cook as district superintendent.

"Another forward step in 1949 was the installation of two-way radio equipment to provide immediate communication between offices and rolling service equipment. This would prove invaluable in expediting both construction and maintenance.

"The long-awaited postwar building program also was started in 1948. Before the year was out, a 33 percent gain in membership would be realized through the addition of 1,182 homes. This gave Warren RECC 4,794 members at the close of 1948.

"With the dawn of a new decade, Warren Rural Electric Cooperative, in 1950, was ready to launch an even more ambitious program than it had seen in the late 40s.

"This was to be a year of growth—a decade of prosperity. New homes were being constructed in large numbers, with the swing being toward urban living away from the more heavily populated areas. And there still were older homes clamoring for electric service.

"A phenomenal 60 percent growth in membership was registered by Warren RECC during 1950 alone, the number of consumers jumping from 6,561 at the close of 1949 to 10,908 for a gain of 4,347.

"Before this new decade was closed, the Cooperative was to see its membership grow steadily each year to a total of 17,010 consumers. And with this growth and expansion of facilities, 98 percent of all homes within the Warren RECC service area would be electrified.

"The 'Dark Ages' which had existed in rural areas for so long was now coming to an end with increasing rapidity. This could well be

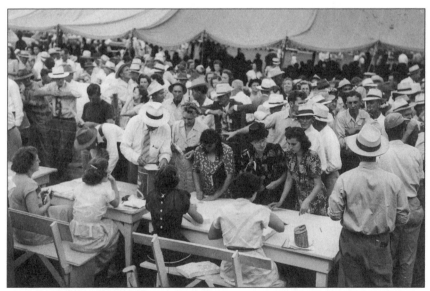

"Sign in please, so we'll know you've been here and your vote will count. And, don't forget to sign up for the door prizes!" For over seventy years, annual co-op meetings have been the highlight of the year for many a rural family. (Warren RECC archives)

The humblest home can have a meter and an electric connection. This was one of the earliest residential services in the Warren RECC area. (Warren RECC archives)

referred to as the 'age of electric lights and power.'"

Rick Carroll, Director of Programs and Communications, has captured the magnitude of the outreach of Warren Rural Electric Cooperative Corporation: "Almost simultaneously with the interest in new industry that would provide more jobs for rural people, Warren RECC became the leader in an effort to supply rural areas with a reliable source of centrally distributed water. This was a luxury that none of our rural areas enjoyed, and it was essential for two major reasons: water, through the drilling of wells, was becoming harder to find and that which was located was most often contaminated. Two, industry could not be enticed into rural areas unless there was an ample supply of pure water available.

"The Richardson community in northern Warren County became the first experiment in establishment of a rural water district. This was accomplished through procurement of a $60,000 FHA loan and a $46,000 grant, and in October 1964, the project was completed.

"Although the number of consumers was comparatively small, it proved that rural water programs could be projected successfully. It also created the atmosphere for construction of new homes, improvement of existing homes through installation of indoor toilet facilities, and raising overall standards of living.

"The cooperative's second step in this direction came with establishment of Northside Water District, which was begun in 1964 with a $350,000 FHA loan and a $205,360 grant. This initial Northside project, located north of Bowling Green, was completed in November 1965.

"Many expansions of Northside District were to follow in ensuing years, and one of these, in 1968, consolidated the Richardsville Water District with Northside District.

"Activities of Northside Water District were essential to the procurement of industry in rural Warren County. Not only did it supply water to that county's second Industrial Park, but it laid lines in sufficient size to serve the needs of both Firestone and Chrysler plants as well as any future industry in those areas.

"Sewage facilities also were a requirement of these industries if they were to locate in rural areas, so this was undertaken by Northside Water District with the cooperation of the city of Bowling Green.

"Northside District, in its sixth major project, undertook the construction of seven miles of sewer line to serve initially eighty consumers, including two schools, numerous residences, and small businesses along with industrial operations.

"To help perpetuate these water districts, Warren RECC not only provided the management and maintenance of these facilities but also assisted through bookkeeping and bill operations.

"Other counties in the Warren RECC service area also were desirous of establishing rural water districts through which their rural people could be served. Through the 1960s, several thousand new jobs had been created through the attraction of new industry. And thousands of rural families were finding that they could remain on the farm or construct new homes in rural sections while enjoying every convenience that could be found in more heavily populated cities."

In a 2005 interview, Charles M. Stewart explained why the statewide

In the early '60s, Warren RECC's first installation of security lights turned nights into days on the J.R. Bettersworth Farm on Three Springs Road. It was even bright enough to work on tractors out in the barn lot. (Warren RECC archives)

association [KAEC] was established: "Private power companies were fighting the rural electrics. It was dog-eat-dog in those times. That's how the state association came about. We had some good leadership. We had a lot of political power and reached a lot of people, and a lot of people wanted electric service who didn't have it."

When J.K. Smith said, "Start with people—it's a people program," he might have had in mind Mrs. Johnnie Cline of Gold City Road in Simpson County. "We lived in the Gold City area before there was any electricity. We lived on a farm, milked cows by hand, had to have lanterns to feed the cows. The lanterns hung in the milk room while we milked. Carried lanterns to other parts of the barn to feed the horses and mules. Had sows and pigs to feed. Would raise the pigs until they got to be hogs around 250 to 350 pounds to dress for our meat for the year.

"When it came wash day we used a large black kettle set up outside to heat the water, boil the clothes. Had to have wood from the woods to heat the water. We put our milk and butter in the cellar where the canned vegetables and fruit were.

"The people that got electric were so proud—sure did change our lifestyle. Now everybody should appreciate Warren Rural Electric Co-op. When we have snow and ice storms, wind, rain, tornadoes—they're out all hours night and day working to get our current back on. Thank God for what we have."

WARREN RURAL ELECTRIC
COOPERATIVE CORPORATION

Miles of Line:	5,571
Consumers billed:	58,556
Wholesale Power Supplier:	TVA
Counties Served:	Barren, Butler, Edmonson, Grayson, Logan, Ohio, Simpson, and Warren

ADMINISTRATION

Lester Wright	1938 – 1948
Ollie Mae Roberts	1948
Charles M. Stewart	1948 – 1980
Floyd H. Ellis	1980 – 2000
Gerald W. Hayes	2000 – 2008
Gary K. Dillard	2008 – present

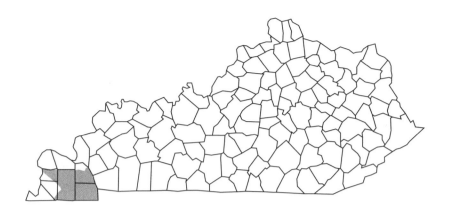

WEST KENTUCKY
RURAL ELECTRIC COOPERATIVE
CORPORATION

"We haven't lost sight of people."

David E. Smart
President and CEO
West Kentucky RECC

David E. Smart takes a step back and listens. He's the President and CEO, but he doesn't interrupt a reporter's conversation with eighty-two-year-old Joe Hargrove, eighty-three-year-old Joe Lovett, eighty-seven-year-old Jeffery Howard, eighty-three-year-old Mary Neale Barton, and ninety-year-old Ralph C. Edrington (Board President) in the Board of Directors room at 1218 West Broadway in Mayfield, Kentucky.

The center point of the Jackson Purchase is also the land of author Bobbie Ann Mason, journalist Joe Cross Creason, and humorist Irvin S. Cobb. In the quiet of the day there's insight into life before electricity, the early days of rural electrification, and what it takes to spend a lifetime working for a rural electric cooperative.

"Life before electricity?"

Ralph Edrington, angular as he is spry and a forty-one-year member and current President of the Board of Directors of West Kentucky RECC: "There were three of us boys in the family. We gathered up all the old wood to fire up the kettle. Monday was wash day. That was our job—to get the fire hot and keep the wood stove going. We helped her [his mother] to do the heavy part. Tuesday was ironing day. I have a 'Labor Saver' washboard. It's in my den now. We had a basement in our house. We stacked ice in it and covered the ice with gunny sacks and sawdust. Made ice cream!"

Mary Neale Barton: "We took the tub out in the sun to heat the water—it was a nice warm bath."

Jeffrey Howard: "We had Delco batteries, big banks of batteries...had a generator... had light.... twenty-eight volt pump...1,000-gallon water tank. We piped out to houses around."

"The early days of rural electrification?"

One of the earliest power station installations and the crew that built it. West Kentucky RECC gets its power from TVA, but the grit it took to get the co-op up and on-line was gleaned from local boys, some of whom were probably still farmers. (West KY RECC archives)

*Thirty years before this photo was taken an electrified farm was only a dream for a rural family, but as the Wilson kids are proudly professing, their farm and home had become totally electric—an Electrofarm! (*West KY RECC archives)

Ralph Edrington: "I'm ninety years old. I was at UK when the lights came on. Graduated in '38. Sometime when I was at UK they turned the lights on in Carlisle County." The point is vividly made—lights on the campus of the University of Kentucky, but not in faraway Carlisle County—a classic case of "haves" and "have-nots."

Joe Hargrove: "Some were skeptical that it [electricity] would work."

Mary Neale: "They were eager—they *wanted* electricity."

Joe Hargrove: "But they'd rather have the pole on the other guy's land."

Ralph Edrington: "I was helping my dad getting right-of-way. We had to go get the sheriff because this one guy wouldn't let us go across. You cannot appreciate what it was like unless you were there." Ralph Edrington was there.

"On working for a rural electric cooperative?"

Mary Neale: "When I came to work in '43, there were 2,500 consumers and 500 miles of line. That was with trucks with no radios."

Joe Hargrove: "I started out as radio maintenance technician, then I graduated from that into meter technician. I spent thirty-two years with West Kentucky RECC."

Joe Lovett: "I went to work with REA in '47 and retired in '87. I'll be eighty-three my birthday. I call him [Joe Hargrove] 'old man.'"

Joe Hargrove: "One day Mr. Lovett and a young man and I were setting some huge transformers. I was up on a platform, and I thought the rail was behind me, but it wasn't. When I fell, I landed on this big, young guy. He said it didn't hurt."

Joe Lovett: "I had a close call close to General Tire. I was doing maintenance on a line and we'd switched out so we could kill the line. A guy called and said that the line was dead, but it wasn't. Caught my clothes on fire, down my arms, arced and I had to scrape the melted nylon material from my shirt off my arms. I was an apprentice at fifty cents an hour. Worked for six months, then I got raised to sixty cents. Worked a year at sixty cents. As a lineman I made ninety cents. Someone's lights were out for two or three days and we'd get a postcard telling us. There was no other way to let us know."

Mary Neale Barton: "There were three women and one man and one manager. [The office] had a Warm Morning stove. We ladies had to go in and build the fire in that stove. First office had a car repair shop on one side and a poolroom on the other."

Joe Lovett: "When I first went to work we used a spade and a spoon to get the loose dirt out. Engineers staked it off, and we had to dig the holes."

Joe Hargrove: "We used mules where you couldn't get in with a truck. Most of the time the owner would take care of the mule. The mule would drag the pole and we'd set it by hand. Set in winter the same way—took an iron pick to break the ground."

The entire eastern service territory of West Kentucky RECC borders Kentucky Lake, one of the world's largest man-made lakes. Kentucky Lake was formed when a dam was constructed on the Tennessee River in Gilbertsville, Kentucky. Kentucky Dam was constructed over a six-year period beginning July 1, 1938. The dam is owned and operated by the Tennessee Valley Authority and has been instrumental for years in controlling flooding in the region on the lower Ohio and Mississippi rivers and for generating electricity. TVA and the construction of Kentucky Dam played a significant role in bringing electricity and a better quality of life to rural western Kentucky.

Ralph Edrington: "The main thing was it livened up the economy in Calloway, Marshall, and Trigg counties. These counties had been among the poorest areas in Kentucky until the dam was built and the power came. All the chemical plants and the jobs that came with them in the area came to life because of TVA and the availability of power."

David Smart: "Kentucky Lake and the Land Between The Lakes region between Kentucky Lake and Lake Barkley are the primary tourist attractions in the area. They draw water sports enthusiasts in addition to many fishermen and hunters. There is no telling what the economic impact has been and continues to be in the area because of Kentucky Lake."

With passage of the TVA Act in May of 1933, the formation of the

Around and around the playground they wound while lining up to register at the annual meeting. There was no time to see-saw here. These folks knew what they wanted, went after it, and wanted to be a part of the decision-making! (West KY RECC archives)

Annual meeting drawings of lucky numbers lead to gifts that brighten days and nights and put the rooster out of business—electric clocks, irons, toasters, percolators, and cookbooks for the electric kitchens back home. (West KY RECC archives)

Rural Electrification Administration in 1935, the passage of the Rural Electrification Act in 1936, and the TVA expansion into rural western Kentucky, the pieces were in place for the formation of electric cooperatives, including West Kentucky RECC.

In April of 1938 a small group of rural leaders from Graves County met at the Stovall building in Mayfield and officially organized the Cooperative for the purpose of making electricity available in the surrounding rural area. At that time, less than three percent of the farms in Kentucky had electricity. West Kentucky RECC received its initial approval from the REA to build 198 miles of line and serve fewer than 400 members in that area.

In a conversation with President and CEO David E. Smart during the latter months of 2007, he shared the following: "My idea would be that the book [*Let There Be Light*] truly emphasize the impact rural electrification has had in Kentucky. Not only from a quality-of-life standpoint, but also from an economical standpoint. Many people believe that the TVA cooperatives are run by TVA, or that TVA is their provider instead of the local cooperative. Therefore, I think that some emphasis should be given to the cooperative business model and the relationship between wholesale provider, transmission provider, and distribution provider. TVA is a generation and transmission provider, West Kentucky RECC is a distribution provider."

David Smart continues: "When I took my master's through Morehead State, there was very little about the cooperative business model. I got a job at Fleming-Mason RECC and worked as engineer. I was manager of engineering for ten years and then West Kentucky had an opening for CEO and I came here."

Smart emphasizes the fundamentals of a cooperative business model: "First, cooperatives are member owned and governed. Second, cooperatives are non-profit. Third, cooperatives were created to serve and meet the needs of their member-owners.

"When I hire new employees I emphasize three key elements of the cooperative way of doing business. First, never lose sight that we are a service organization. Second, do whatever it takes to get the job done in a safe and efficient manner. Third, the co-op is based on Biblical values, including honesty and integrity, and this is major.

"The electric cooperative industry still has Biblical values as its base. It's nothing to go to a national meeting where they start the meeting with prayer. The industry is full of good Christian people, who have been given direction, who have bought into the program and who get the job done.

"Employees in the cooperative world are like a family. We work together for the betterment of life and the quality of life in the small towns in which we live and serve. Cooperatives are a good place to work.

"The Golden Rule goes a long way in how you treat people."

"The future?"

"It's bright. Co-ops are in a position to grow even larger and stronger in the next few years. We are not in a very commercialized area, but we're agriculturally driven. Areas are changing from 100 people farming 500 acres to just a few farming the same amount of land."

"Leadership?"

"The key to being successful as a leader, you have to learn each individual's personality. Have to learn what is a passion to them and what their strengths are and put them in a position to use their strengths to accomplish whatever you need for them to do.

"The biggest hurdle to the area is that West Kentucky buys wholesale power exclusively from TVA. TVA sets the wholesale price and regulates the retail rate. TVA is the high-cost wholesale provider in Kentucky, but it has a lot fewer requirements as a regulator than the Kentucky Public Service Commission."

"Your greatest challenge?"

"Our greatest challenge is trying to maintain affordable electricity and great service…while at the same time, dealing with escalating costs—wholesale electricity prices—material prices – copper and steel have gone up hundreds of percents…and maintenance costs. You can't let the system go. How do you provide good efficient service, a good employee package, a good work environment, while at the same time dealing with escalating costs—not only from a material standpoint, but from an employee standpoint as well?"

From its humble beginnings in 1938, West Kentucky RECC has grown tremendously in its first seventy years to serve more than 31,000 members with more than 37,500 meters across nearly 4,000 miles of line in Calloway, Carlisle, Graves, Hickman, Livingston, and Marshall counties.

The day of the kerosene lamp, the wood cook stove, and the washboard has passed, thanks to the efforts of that small group of rural people working together for their own and the common good to make a better life for themselves and their neighbors.

Imogene Vick remembers: "My family moved to a ninety-acre farm in Calloway County about 1936. My grandfather lived alone and needed someone to be with him all the time. This farm was located about eight miles east of Hazel, Kentucky, and about two miles off the Tennessee line.

From its 1938 beginning, to its second home from 1942 to 1952 at East Broadway and North 5th Street, West Kentucky RECC has grown from its original 198 miles of line with fewer than 400 members to almost 4,000 miles of line and 38,000 members. (West KY RECC archives)

"There was a two-room house with a kitchen and back porch that had been added to the two front rooms. A huge fireplace was our only source of heat and a wood cook stove was in the kitchen. We had kerosene lamps for our light and later we had an Aladdin lamp, which gave out a brighter, white light.

"I remember my folks talking about our part of the country getting electricity. The electric lines began to be built all around us it seemed, but our place was out of the way the lines were being built, so the lines came across the county on both sides of our farm. My father worked to get a line to our house but time went on and nothing happened.

"About 1946, my father decided to tear down the old house and use the lumber and sell some trees off the property to build a better house. He decided to go ahead and wire the new house in hopes that someday he would get electricity. Finally, about 1948, the power company ran a line across country through our fields and woods to hook us up to the lines. It was a happy time to finally have electric lights, and soon we bought a refrigerator, which kept on faithfully running until my father's death in 1982.

"I remember I could hardly wait to get to town to buy an electric iron. In fact, I went to the field where my father was plowing to ask permission to go right then to get the iron. How wonderful not to have to put the heavy irons on the stove to heat and keep changing them out as they cooled. Today how much we take for granted the wonderful convenience that we have in the electric power in our homes."

WEST KENTUCKY
RURAL ELECTRIC COOPERATIVE CORPORATION

Miles of Line:	3,970
Consumers billed:	37,685
Wholesale Power Supplier:	TVA
Counties Served:	Calloway, Carlisle, Graves, Hickman, Livingston, and Marshall

ADMINISTRATION

Robert Usrey	1938 – 1942
H.E. Pentecost	1942
J.C. Roby	1942 – 1944
George Knight	1944 – 1948
John Edd Walker	1948 – 1986
Michael Alderdice	1986 – 2001
Ron Mays	2001 – 2002
David Smart, P.E.	2002 – present

KENTUCKY ASSOCIATION OF ELECTRIC COOPERATIVES
(KAEC)

History is of little value unless we learn from it.
And what better way to learn than to document, practice what we
learn and put what we learn to good use. And that's why Kentucky's
electric co-ops chose to have this book, Let There Be Light, *written.*
Many of the earliest pioneers are passing away and this
needed to be done now.
And so have we committed ourselves to this task.

Ronald Lee Sheets, President
Kentucky Association
of Electric Cooperatives

From the archives of the Kentucky Association of Electric Cooperatives comes the following history of the association, authored by J.K. Smith, Ron Sheets, and their staffs:

"In December 1937, Kentucky's first REA-financed lines were energized by Kentucky's first rural electric co-op, Henderson County REA (later consolidated with Union County RECC to become Henderson-Union RECC and subsequently, along with Green River

289

Electric Corporation, Kenergy Corp.).

"The idea for the statewide organization, KRECC [Kentucky Rural Electric Cooperatives Corporation], began to take definite form at a meeting of state co-op leaders in Shelbyville in 1941. State co-op leaders began to realize the need for a statewide headquarters, owned and controlled by the Kentucky co-ops, to serve as their official organization. Such an organization was needed to spearhead the battles against anti-cooperative interests. Individually, the co-ops could not command the same respect or be in position to mold favorable public opinion. However, acting collectively through this proposed central organization, they could become strong and rally favorable public opinion toward their cause.

"August 26, 1943, about eight years after the creation of the Rural Electrification Administration as a government agency, the Articles of Incorporation of the Kentucky Rural Electric Cooperative Corporation were officially recorded and its first Board of Directors appointed until there could be a regular election of Board members. These first Board members were: C.E. Miller, Brandenburg; Robinson Cook, Danville; D.W. Howard, Pineville; H.W. Daniels, Owensboro; and G.L. Bridwell, Cynthiana. However, mainly due to World War II, the state association remained an organization in name only, without a full-time staff, until January 1, 1948. At a meeting in September 1947, held at the Seelbach Hotel in Louisville, the Board members had voted unanimously that a full-time staff should be employed to operate the state office and carry on other KRECC activities.

"Three months later, James K. Smith, who had been the Manager of the Fleming-Mason Co-op at Flemingsburg, became the state association's first executive manager. Employed for the purpose of establishing a permanent statewide cooperative office, he rented a two-story, seven-room frame house at 1412 Bardstown Road in Louisville for $75 per month. This became the association's first home and headquarters. His first two employees were Mrs. Nell Ashcraft, who had been his secretary at the Fleming-Mason Co-op and had come to Louisville to serve in this same capacity, and Miss Ann Jones, formerly the office manager at the Harrison County Co-op at Cynthiana, who was employed as the KRECC office manager.

"The Accounting-Central Billing Department shared the first floor of the office with J.K. Smith and his secretary, Mrs. Ashcraft. Most of the second floor was used as storage space until February, when Seth Thompson, who had been working with the U.S. Department of Agriculture Information Services in Washington D.C., was employed as the first editor of the KRECC newspaper, *Kentucky Electric Co-op News*. This newspaper was another of the first services offered by KRECC to the members of member co-ops. Its prime purpose was to keep the co-op members informed about their programs, REA news and activities, and to provide certain educational material that might be helpful to the farm consumers in improving rural living.

"The second-floor, semi-attic office of the first newspaper staff got rather hot in warm weather, and Seth Thompson literally sweated out his first issue, which was published in April 1948. It was an eight-page, 17 1/2 x 12-inch, five-column newspaper, liberally sprinkled with pictures, containing national and local co-op news and some helpful hints for the farm housewife. In June 1948, Louis Miller was employed as advertising manager. *Kentucky Electric Co-op News* was published monthly and distributed to co-op members for thirty-five cents per year subscription.

"KRECC also began handling to some extent the sales of line

From late 1947 to mid-1959, KAEC had three offices, all three of which the organization outgrew until they moved into the current Bishop Lane location. Here the office staff poses in front of their new location which, in 2009, will celebrate fifty years of housing offices, warehouse, and shop. (KAEC archives)

equipment to the co-ops, one of the state association's first services to the co-ops.

"Still another service offered from the very beginning was a safety and job training program. This program was administered by a committee composed of three managers and three co-op linemen and directed by a full-time safety and job training director. The first director for the state association was E.A. Reid. He constantly traveled the state, visiting the co-ops and assisting in the promotion of their safety and job training programs.

"The Frankfort law firm of Ardery and Pritchard was retained to handle KRECC legal matters.

"In June 1949, the KRECC operation was still expanding. A youthful, rapidly progressing organization, it was fast outgrowing its present facilities. Additional space was needed for increased personnel and services. To meet this need, J.K. Smith rented six rooms of office space at 3716 Lexington Road in St. Matthews on the second floor over some stores. These had formerly been the state offices of the old Kentucky Farm Bureau.

"About this time, a new job of full-time field representative was set up to facilitate the obtaining of line materials in quantity and selling them to the co-ops. Charles Foley, the Sales Manager of the Statewide Electric Corporation, was employed to handle this expanding service.

"In October 1949, KRECC set up a print shop in some rented space in the Lewis Seed Co. on Spring St. as an added service to the co-ops, printing letterheads, forms, etc., for them at reduced cost.

"In January 1950 some important changes took place in the KRECC publication. The newspaper format was changed to a slick-cover, 8-1/2 x 11 inch, 16-page magazine with a colored cover and localized inserts for those co-ops requesting this service. This magazine format had been planned for months in order to bring co-op members the finest publication possible. The circulation was now 94,062 copies monthly.

"The state association was still experiencing growing pains and again found itself becoming cramped in its present facilities. Plans were already being made for the addition of a new Special Services Department to repair transformers and meters for member co-ops. There just wasn't enough room at the Lexington Road office, so J.K. Smith made plans to move again.

"In September 1951, a former farm equipment warehouse at 1430 Mellwood Ave. was purchased for $80,000 to meet the increased need for space due to expanding facilities and increased personnel.

"The third home of KRECC in its progressive expansion was the largest yet. The large first floor and basement seemed to

Beauty contests held at practically every rural electric cooperative at annual meetings, especially in the 1960s, resulted in many a young lady advancing to the Miss Kentucky pageant. Here's an example of some of Kentucky's loveliest local RECC winners as they prepare for their walks down the runway. (KAEC archives)

provide sufficient space for any anticipated expansion at that time.

"The Engineering Department was created to provide engineering services to the member co-ops, and the first engineer, Herman Williams, was employed.

"In December 1951, the plans for the new Special Services Department became a reality, and this department began operating in a large section of the basement. Wallace Rea, the first head of this department, and one helper, repaired transformers while Ed Bryan was employed to repair the meters.

"About a year later, in 1952, the Member Services Department was added to KRECC as a further service to assist member co-ops in their power use programs. Ervin R. Baker was employed as the first agricultural engineer to head this new department.

"Also, in February 1952 the name of the KRECC magazine was changed to *The Rural Kentuckian*, and later to *Kentucky Living* in April 1989. In May 1952, the Special Services Department added an oil filtering service for co-op transformers.

"In 1954, the Consumers Credit RECC, a separate cooperative working in conjunction with KRECC, became a part of the KRECC operation to financially assist members of the co-ops to obtain electric equipment and appliances at a cost that they could afford. Ken Herren became its first director. His staff originally consisted of one employee, but was shortly increased to two employees. In September of this same year the Special Services Department added another service to the co-ops. Linemen's rubber gloves were tested for possible defects or leakage—a service that still continues by contract through a third party.

"In 1955 a new job was added to the Member Services Department to further extend its services to the co-ops. Mary Alice Willis was employed to fill this new position of home economist to promote power use in the home.

"In January 1957, with the increased activity in assembling transformers for the co-ops, the Special Services Department added a conveyor belt system to facilitate the transformer assembly.

"In July 1958, a new department was created to combine and centralize engineering, purchasing, and warehouse activities.

"Beginning in January 1960, the Statewide Rural Electric Cooperative Corporation was formed, which became the trade name for materials marketing activities. Later, in December, a joint agreement was developed between this organization and a related materials program in Ohio known as RESCO. Later, in 1963, the organization was expanded into the state of Illinois and in subsequent years was extended into a number of states east of the Mississippi River with warehouses being strategically located in various parts of the country.

"This continued KRECC expansion of services and personnel to handle these services began to fill up the state association's third home, and increasingly cramped quarters caused J.K. Smith to begin thinking about the necessity for a new and larger building as early as 1955.

Tentative plans got under way and finally began to take more definite form. Possible building sites were examined. Finally, on November 6, 1958, groundbreaking ceremonies were held on a building site that had been purchased on Bishop Lane near the Watterson Expressway in Louisville.

An inside view of the transformers that are manufactured and marketed by United Utility Supply, an affiliate of KAEC since 1957. (KAEC archives)

Construction was begun on a new building to house offices, shop, and warehouse facilities. This building was occupied in 1959 and remains the state headquarters.

"In June 1959, the personnel of the Kentucky Rural Electric Cooperative Corporation occupied its brand new $500,000 home on Bishop Lane.

"This new building, containing office, warehouse, and shop facilities, became the fourth home of the state association.

"The building's location on a six-acre site just off the Watterson Expressway makes it easily accessible to visitors from all parts of the state.

"This new building was made possible through the combined financial support of twenty-one Kentucky co-ops who contributed a total of $510,000 toward its construction.

"As KRECC has grown and expanded, so have the state co-ops. Practically all Kentucky farms now have electricity as opposed to about three percent when REA was first established in 1935. Today, the co-ops are now serving thirty-five percent of Kentucky's population with more than 90,000 miles of electric lines.

"J.K. Smith's vision included United Utility Supply, a material supply organization owned by more than 200 member cooperatives in nineteen states; Central Area Data Processing, a co-op based in St. Louis that serves the data processing needs of hundreds of rural electric cooperatives; and the National Rural Utilities Cooperative Finance Corporation (CFC), an institution that provides capital to the nation's rural electric cooperatives to supplement loans from the Rural Electrification Administration.

Louis B. Strong, second president of KAEC. "A polished western gentleman— a man with a warm embracing style." (KAEC archives)

"J.K. Smith served as interim President and General Manager for East Kentucky Power until Don Norris was hired. Smith was part of the original group that organized East Kentucky in 1941. When United Utility Supply became an affiliate of KRECC in 1957, it became the first statewide marketing entity in the nation, as KRECC became the first transformer-manufacturing cooperative. Also while at KRECC, Smith helped start Central Area Data Processing, which now services fifteen million accounts for rural electrics in fifteen states. (East Kentucky Power's Shanda Crosby interview.)

"On August 5, 1970, after almost a quarter century of service, J.K. Smith announced his resignation from KRECC to become Governor and CEO of the National Rural Utilities Cooperative Finance Corporation (CFC) in Washington, D.C. His official resignation from the state association became effective October 31, 1970.

"Finding a person who could step into the shoes of J.K. Smith was no small task. The association's leaders looked nationwide for the best person. Ultimately, the leadership settled on a gentleman from Blackwell, Oklahoma, Louis B. Strong. He was the General Manager of the Kay Electric Cooperative and was serving on the Board of Directors of the National Association of Electric Cooperatives (NRECA) in addition to his responsibilities at the Oklahoma electric co-op.

"Upon accepting the position of President of the Kentucky Rural

Electric Cooperative in Louisville (December 1, 1970), an arrangement was made to permit Louis Strong to continue serving on the Board of Directors of the national association. He concluded his service in the officer chairs at NRECA by being elected Chairman of the Board, a position he held for the years of 1971 and 1972. He remained on the NRECA Board until his successor was chosen.

"Louis Strong was held in high regard nationally for his service, not only as the top officer of the national association, but by his related service on a number of energy study committees at the national level. A polished western gentleman, a man of significant character, and one with a warm embracing style, Strong served well during his tenure with the state association.

"Perhaps Louis Strong's most significant contribution of his many accomplishments while at KRECC was the passage of one of the strongest utility territorial bills in the United States, enacted by the Kentucky General Assembly in 1972. Its passage involved literally years of effort and research and was the product of both Strong's leadership and a highly-supportive governor in the person of Wendell Ford, together with the expert political judgment of the Green River Co-op manager in Owensboro, J.R. Miller.

"It was during Louis Strong's tenure that the Kentucky Rural Electric Cooperatives Corporation Board of Directors voted to amend the Articles of Incorporation to change the name of the corporation to Kentucky Association of Electric Cooperatives [KAEC]. The vote was taken June 18, 1974, and filed with the Kentucky Secretary of State on August 5, 1974.

"Upon his retirement at the end of 1981, Louis Strong did some consulting work on industry matters for a few years before permanent retirement. He and his wife, Eileen, continued to live their retired years in the Louisville area. Both are now deceased.

"The association chose its third president at a special board meeting in August of 1981 by selecting Ronald Lee Sheets of Frankfort, Kentucky, who was currently serving as Vice President of Government Relations for KAEC. Prior to coming to KAEC in 1978, Ron had ten years experience working with the Kentucky General Assembly as Vice

President of the Kentucky Retail Federation—a state association representing retail merchants and headquartered in Georgetown.

"One of Ron Sheets' driving efforts was to see that all electric co-ops in Kentucky eventually became members of the state association. At the time of his hiring, four systems were not association members—Salt River of Bardstown, Taylor County of Campbellsville, West Kentucky, headquartered in Mayfield, and Pennyrile at Hopkinsville. Within five years, all of the co-ops became full dues-paying members of the association. Each one of Kentucky's twenty-six co-ops remains a member of the state association, a reality Mr. Sheets notes as his greatest satisfaction.

"Another of his efforts, which began in 1982, was to establish the highest award recognized by Kentucky's electric cooperatives—the Distinguished Rural Kentuckian Award. The first such recognition was provided to the internationally known poet Jesse Stuart of eastern Kentucky, followed the next year by longtime and highly regarded U.S. Senator John Sherman Cooper, who was also appointed by President Kennedy as Ambassador to India following his tenure in the U.S. Senate.

"One of Ron Sheets' commitments has been his personal presence at co-op annual meetings during the summer. Upon his intended retirement by the end of 2009, he will have attended more than 700 of these meetings over his thirty-one years of employment, probably a national record. He continues to treasure the opportunity to come in contact with what former Governor 'Happy' Chandler often described as the salt of the earth. That's what local co-op members are throughout the Commonwealth of Kentucky—the salt of the earth.

"Ron Sheets served one term on the board of directors of the National Association of Electric Cooperatives and two terms as President of Kentucky Friends of 4-H.

"In December 1977, the board approved plans for an additional expansion, which allowed the manufacturing of tanks into which the KAEC-manufactured transformer is placed, along with approval for additional warehousing space.

"In July 1996, the organization produced its one-millionth transformer.

"In December 1999, additional warehouse space was purchased in

Louisville, the property being acquired from the ENRO shirt factory—very close to the KAEC headquarters. All of the warehousing operations were moved to this location, currently named the Cooperative Distribution Centre. A completely new transformer painting center became operational in mid-2008.

Ronald Lee Sheets, third president of KAEC. "History is of little value unless we learn from it." (KAEC archives)

"The entry of Kentucky into the transformer manufacturing business was made necessary since the material marketing organization was cut off as a distributor for Maloney Transformers in 1956. At the time, Maloney was the largest distributor for the transformers in the South.

"Internal historical department reports document from an August 23, 1968, internal memo that the entry of Kentucky into the transformer manufacturing business caused competitors to reduce their prices forty-seven percent.

"The mission of the Kentucky Association of Electric Cooperatives is 'to efficiently and effectively represent the collective interests of electric cooperatives in Kentucky.'

"The theme of Kentucky's electric cooperatives is 'Our Power Is Our People.' It's a phrase that was adopted in 1982, and is also the title of a print the association sponsored in 1984 when it commissioned Michael Finnel of Frankfort to travel the state and randomly select individuals to pose for inclusion into the painting from which the prints were developed. The individuals are assembled in the geographic configuration of the state's borders with the soft urban lights of a city in the distance. The original painting is hung at the association's headquarters building in Louisville but was also on display at the front reception desk at the State Capitol for a period of time.

"The association has initiated many efforts over the years that have served to improve the quality of life for almost everyone in Kentucky. A prime example goes back to the 1950s when J.K. Smith, together with leadership from the University of Kentucky College of Agriculture, began discussions of a trade show for farm machinery equipment. It was believed that the show could serve to better educate the farm population in Kentucky as to both the availability and the use of the newer components of farm equipment then available in Kentucky. This began in the corner of a tobacco warehouse in Lexington and rotated to various regions in Kentucky through the University of Kentucky College of Agriculture on an annual basis before becoming permanently housed at the Kentucky Fair and Exposition Center in Louisville. This has led to what is known today as the National Farm Machinery Show—the largest activity of its type in the world, and it began with an initiative from the electric co-ops in Kentucky and the University of Kentucky College of Agriculture. It was a marvelous concept that has resulted in a tremendous success.

"Another example of the collective efforts of the co-ops resulting in public good is the Safety and Loss Prevention Programs. Numerous times over the years, our co-op linemen or other employees trained in CPR techniques have come across individuals who have almost drowned, suffered significant personal injuries, etc. Through immediate lifesaving efforts, the assistance lent to these people through co-op safety and loss prevention training has saved many lives.

"The statewide magazine for Kentucky's electric co-ops, *Kentucky Living*, is distributed monthly to approximately one-half million people, and has the largest monthly circulation of any magazine in the Commonwealth. Each subscribing co-op has a local feature section highlighting detailed local information to support the statewide information of

From a drawing of wire, a plug, and a socket by Drew McLay emerged the mascot of NRECA — Willie Wired Hand—who was touted in the 1950s as the new hired hand that brought electricity to farm families. (KAEC archives)

A capital tour! In the early 1960s, President Lyndon Johnson proposed that the cooperatives send students to Washington, D.C., to "see what the flag stands for." In 2004, the Kentucky contingent was part of more than 1,200 students from all over the U.S. to visit the nation's capital. (KAEC archives)

the magazine.

"Rural Cooperatives Credit Union is a member-owned financial cooperative that was chartered in December 1964 for the purpose of serving employees and family members of the Kentucky rural electric cooperatives and member rural electric cooperatives. Eleven people joined together to charter this credit union under the state laws in Kentucky, and since its formation it has grown noticeably. This growth is visible today as this financial cooperative provides services to more than 6,000 members in several states. Also, the evolution of Rural Cooperatives Credit Union is apparent in the change from a one-person office with part-time hours in the early years, to today's operation of a full-service financial institution with eighteen staff members and two offices open five days a week. This credit union has grown because of the direction received through the years from its Board of Directors and other volunteers. Today there are twenty-three Directors and a four-member Supervisory Committee involved in shaping this credit union's future.

"Rural Cooperatives Credit Union primarily provides consumer savings and loan services to its members. The National Credit Union Administration, a U.S. government agency, insures it with its deposits to at least $100,000. The credit union's mission today is to promote thrift and provide the best possible financial services to all members at competitive rates, while maintaining the credit union's long-term financial stability.

"Underlying the very idea of credit unions is a fundamental set of values shared with other types of cooperatives. The formation and growth of this credit union has been fashioned from the cooperative principles that are key in the history of years. The vision of many volunteer directors and committee members must be acknowledged when looking at the successful growth and operation of the Rural Cooperatives Credit Union."

Kentucky Association of Electric Cooperatives has grown tall from the seed planted by J.K. Smith. Without his imagination and persistence, along with many other emerging rural leaders, "Let there be light" might have remained words on the pages of threadbare family Bibles.

A signal sign of KAEC's commitment to community and the youth destined to become its future leaders is the annual Kentucky Rural Electric Washington Youth Tour. It's a weeklong, all-expenses-paid trip to the nation's capital. The tour is open to select high school students whose homes are served by electric cooperatves. The mission of the tour includes: recognition and reward for academic achievement and community leadership; fostering students' appreciation for the democratic form of government; and building students' leadership skills so that they may make a difference in their communities.

One participant, Phil Bozarth, representing Kenergy Corp., was seventeen years old when he went on the tour. He summed up the experience: "Co-ops are the essence of the American Spirit."

A student wishing to be considered for the next Washington Youth Tour (which includes a one-day tour of Frankfort) must be a high school junior whose parents or guardians are electric co-op members. Participating electric cooperatives publish applications in their local sections of *Kentucky Living* magazine.

AFTERWORD

J.K. Smith
Western Conference of Public Service Commissions
Meeting in Alaska
June 1, 1970

The meeting was attended by approximately 350 representatives of the Public Service Commissions in approximately twelve western states, plus representatives of investor-owned utilities—gas companies, railroads, and a few rural electric representatives.

Vision of a Leader

We are living in a period of rapid change. We must accept it and whenever possible, lead it. If we are sincere in our efforts to move with the tide of change, we may be required to change some of our operating philosophies and to lay aside some of our cherished traditions and concepts.

J.K. Smith

"Probably never before in its eighty-two-year history has the electric power industry been called upon to face more critical and urgent challenges than those confronting it today. These are challenges which face every segment of the industry, regardless of whether they are investor-owned, municipally-owned, cooperative-owned, or federally-owned.

"Unfortunately, none of us can afford the luxury of a casual consideration of these problems. They are not something that is going to happen some time within the next decade...something that will affect only a few isolated areas. These are extremely urgent problems...and they are with us. These challenges have an effect that will cause repercussions in every segment of our society. We must have the determination and creativeness to develop the solutions.

"These challenges may be divided into four categories, each equally urgent and critical:

(1) a threatened shortage of power supply;

(2) a growing shortage of fuel;

(3) numerous financing problems—money costs that make it increasingly difficult for the utilities to hold the line on power rates;

(4) and an apparent collision course on the part of the utilities on the one hand and the environmentalists and conservationists on the other.

"No single segment of the power industry must permit itself to sit back and wait for others to find solutions.

"The American people are demanding that we meet their needs for power. They will settle for nothing less.

"There has always been an implied partnership between the power industry and the general public. In return for certain specific privileges, the utilities must be willing to accept a certain amount of responsibility to the public. We must not hold back the facts, but let the people know about our problems and their possible effects. We need to discuss these issues openly and frankly. The responsibility for facing up to these problems falls upon every segment of the electric industry, including our regulatory agencies as well as all operating groups.

"We are not dealing with a commodity that can be provided or withheld as we choose. The security and safety of our nation and its people depend upon an adequate and continuing supply of electricity. Only food, shelter, air, and water outrank it in importance.

"In the face of a threatened power shortage today, the total demands for power are continuing to increase. They show no signs of leveling off. For example, power demands increased by nine percent in 1969. At this rate, total demand will double in about eight years or less. This means additional heavy investments for facilities to meet this demand.

"The electric utility industry is already the nation's number one industry in terms of capital investment at approximately $100-billion.

"Projections indicate that an additional capital investment of about $350-billion will be required by 1990. Imagine if you will, all this investment coming at a time when interest rates have already reached a level of about nine percent and threaten to go even higher.

"Yet, these investments must be made if we are to meet our growing power demands. The Office of Science and Technology estimates that about 255 new generating plants will be needed by 1990. These plants will have a capacity of about one million megawatts, or three times the capacity of the 3,000 plants now in existence.

"It is difficult to imagine the chaos that would result if large segments of our country were suddenly to be without electricity. In a nation where electric power turns the wheels of industry; does the agricultural processing; provides the energy to entertain us, warms us, cooks for us, cools us, and protects our health, a power blackout for a large portion of our population is unthinkable. When we consider what would happen as the result of such a disaster, we can have a new appreciation for the responsibility we have for serving the welfare of the public.

"Many approaches and concepts have been suggested as ways of circumventing these problems...some workable and some completely illogical. However, we may be coming to that time in our history when we must give consideration to some very basic changes in the framework of the industry. Perhaps we should give some thought to our total power network and consider some new concepts from the standpoint of the operation of the generation and large transmission grids. This may become necessary for two basic reasons: (1) to insure reliability of service, and (2) because of economic necessity.

"During this period when the utilities are being called upon to make the heaviest investments in facilities, they are also being forced to deal with fixed costs which are reaching new highs almost daily. For example, coal is in exceedingly short supply and the prices on this essential ingredient in power generation are steadily rising.

"Even though coal is our most abundant natural resource, many power generating plants are reporting that their supplies are reaching new lows. An article in a recent issue of *Business Week* magazine pointed out that the 1969 coal production was substantially below the 1968 level, and was estimated to fall short of demand by about 16-million tons.

"Electric utilities burn up more than half the coal mined in the United States each year. The critical shortage in this essential resource has resulted in several of the larger utilities purchasing mining facilities in order to assure themselves of an adequate coal supply. While this may provide a partial answer for those who are large enough to take this approach, what is going to happen to the smaller utilities that do not have the financial capabilities to take this route?

"The law of 'supply and demand' is working in the coal industry. Coal is in short supply and the prices are going up. Foreign competition for domestic coal is certainly affecting both the cost and the supply. Foreign markets willing to pay premium prices are forcing costs upward. If the utilities must pay more for this essential commodity, it necessarily follows that the price of the kilowatt delivered to the consumer is going to be affected.

"This is only one of the many fixed costs that are increasing rapidly as the result of pressures exerted by our inflationary economy. We could just as easily talk about the cost of labor, equipment, materials, and countless other expenditures over which we have no control...expend-itures which ultimately must affect the rate the consumer must pay.

"All of us have a very direct responsibility in the area of environmental control and conservation. While some of the more emotional environmentalists may be radical in their demands, there is no denying that there are millions of Americans who are genuinely concerned about this problem. We cannot minimize the seriousness of this situation and brush it aside. As responsible businessmen concerned with the welfare of our nation, we must do our part to find practical solutions. At the moment, it would seem that we are on a collision course with the environmentalists.

"Some proponents of environmental control refused to compromise or cooperate in reaching sound conclusions. They are 'against,' but never 'for' any solution. This trend is evidenced by a headline that appeared in one of our large metropolitan newspapers. This headline read 'Power Plants Damned at Any Site.' Yet, these same people would be just as vehement in their objections if power to meet their needs was not available.

"We must do all that we can to alleviate a very threatening situation, but we must do so in the context of reason and commonsense.

"Some of the conditions that exist in our environment were aptly described in the April 21, 1970, issue of *Look* magazine. I quote from the article: 'This marks a time of warning. We are fouling our streams, lakes, marshes. The sea is next. We are burying ourselves under 7-million scrapped cars, 30-million tons of waste paper, 48-billion discarded cans, and 28-billion bottles and jars a year. A million tons more of garbage pile up each day. The air we breathe circles the earth 40 times a year, and America contributes 140-million tons of pollutants: 90-million from cars—we burn more gasoline than the rest of the world combined.'

"If we were to make one assumption from these statistics, it would have to be that we are in danger of destroying ourselves in our own waste. This is a serious problem and it is urgent. Each of us here today has a very definite responsibility in helping to solve it.

"In our approach to this and to all our problems, all segments of the utility industry must work together. We should and we must work together and act together to promote the general welfare of all the people. There is room for us all...and there are challenges aplenty for us all.

"While everything we have mentioned thus far may seem to be the exclusive problems of the operating utilities, this is not so. Our regulatory agencies are also deeply involved. As representatives of the people charged with the responsibility of protecting the public interest, our state and federal commissions are going to be called upon to accept increased burdens. Since in philosophy and concept, the commissions serve as the advocates of the people, they will feel the first blast of any public indignation or alarm that might come.

"The very same inflationary pressures that have caused the food, the clothing, the medical services, and the many other things we need to live to soar to new high prices are causing the power industry to take a new look at its operations. Without doubt, many rate adjustments will have to be made in order to meet these rising costs. However, we must be fair and honest in our solutions. If any of us are guilty of using this period of crisis as an opportunity to inflate our own profits beyond acceptable limits, then we shall be guilty of abdicating our responsibilities.

"We must not compromise our own integrity or the integrity of the industry we represent. We must not abuse public concern because if these problems are to be solved, we shall require maximum public support and cooperation.

"Now, for a moment, let's turn our attention to that segment of the industry that I represent...the rural electric systems. The approximately 978 rural electric systems in this nation are facing the same difficulties that confront all other segments of the industry.

"Power demands by our consumer-members are increasing at an unprecedented rate—about twelve percent annually. At the present time, we are delivering central station electric service to some 24-million people located in forty-six states. The rural electrics are operating in all or parts of approximately 2,600 of the nation's 3,100 counties.

"In the thirty-five years the rural electrification program has been in existence, the rural electrics have constructed a plant valued at about $7-billion. Our financing has come exclusively from the federal government in the form of loans from the Rural Electrification Administration. These loans must be paid back with interest, and we are proud of the credit rating the rural electrics have established. And needless to say, we are just as proud of the job we have done in bringing electricity to formerly unserved rural areas.

"Every rural electric system has certain unique handicaps to overcome...handicaps that affect our operations and that require heavy investments per consumer in order to provide service. Low consumer density, low revenue per mile of line, and low load factors combine with heavy investments to make efficient operations essential.

"While the rural electrics face many challenges in the years ahead, none is greater or more urgent than the challenge of finding adequate financing to meet our growth requirements. Projections indicate that we shall need approximately $15- to $18-billion within the next fifteen years for new plant investments. As a practical matter, we realize that we cannot expect Congress to appropriate all the capital that we will require.

"Congress has been appropriating...only about one-half of the actual needs of our program at the present time, based upon present loan fund applications filed with REA. Our studies indicate that the need for additional financing is going to continue to grow throughout the foreseeable future.

"In view of these facts, we in the rural electrification program realized some time ago that the funds appropriated by Congress would have to be supplemented with additional financing if we were to meet our growth demands.

"After about two years of almost constant study, analysis, and consultation, we have developed a plan for providing additional financing for rural electrification on a supplemental basis. We emphasize the word "supplemental" because the purpose of the new financial program is to supplement the REA loan program—not replace it. Utilizing the self-help approach, we shall be doing as much as our capabilities permit through the development of our own credit institution. This independent, private institution is known as the "National Rural Utilities Cooperative Finance Corporation" or 'CFC' for short.

"CFC will be operated and controlled by its member rural electric systems, each of which will be represented on the twenty-two member Board of Directors. Representing all geographic regions of the country, the Board members will be elected from the participating systems.

"The Board will employ a Governor who will become the chief administrative head of the institution. The Governor will then employ a staff to assist him in the operations of the institution. Offices of CFC will be located in Washington, D.C..

"Through the purchase of Capital Term Certificates over a 15-year period, the participating rural electrics will provide the necessary equity capitalization for CFC. This equity capital can then be loaned to the member systems. Mortgages obtained from the member systems in return for CFC loans will have a status that will enable CFC to offer them to the private money market.

"As you know, REA currently holds a mortgage on all the property of each rural electric system that has a loan outstanding. Therefore, it will be necessary for REA to accommodate these mortgages in order to permit CFC to market the mortgages it obtains from the systems. In order to accomplish this, the accommodation must provide CFC mortgages with equal status on a ratio basis with those held by REA. This will have to be done on an individual system basis after a feasibility study of the system has been completed. The purpose of this study will be to assure that the CFC loan will in no way jeopardize the security of the government's investment in the system.

"It is doubtful that many, if any, systems can afford to pay the going interest rate charged on the open money market. However, a blend of interest rates between those charged by REA and those of CFC would put the CFC loans within the financial capabilities of most systems. This blended interest rate will be based upon the amount of money—that is, the proportion of the total loan—received from REA and CFC.

"The fundamental principle behind this concept is a belief on our part that each system should pay an interest rate in accordance with its ability to pay. Those who can afford to pay a higher rate of interest should do so. The financially weak systems will require more REA loan funds than those that have greater financial strength. The stronger rural electrics will be required to borrow a greater proportion of CFC funds. Under this blending arrangement of loan funds and interest rates the feasibility of the system will in no way be jeopardized.

"Since the interest rate on CFC loans will be based upon the cost of money on the private money market, we recognize that CFC interest rates will be considerably higher than those on REA loans. However, we believe that through the blending arrangement we have just described, our new financing program will be feasible and workable.

"We have been most encouraged by the progress we have made to date. More than seventy-five percent of the rural electric systems in the nation have applied for membership in CFC. We hope to have this new credit program in operation within approximately 90 to 120 days. Of course, CFC will have a limited loan fund capacity during its first few years of operation, but as we become more seasoned in the private money market, its lending capabilities should increase sharply.

"This new source of financing should assure the rural electric systems of a continuing source of capital. It most certainly will have a significant effect upon our method of operations. In reality, this plan charts a new course for rural electrification...it sets forth a new blueprint for our future. We believe that it will earn a new respect and acceptance for rural electrification on the part of every segment of the electric industry. We want to be, and intend to be, a full participating partner in the industry.

"The many formidable challenges facing rural electrification in this new decade can be divided into three basic categories: (1) financing; (2) power supply, and (3) maintaining the territorial integrity of our systems. All of these requirements are interdependent, each having a very direct bearing upon our ability to meet our full utility responsibilities in the areas we serve.

"We accept these challenges with a full determination to serve our members' needs and to live up to our corporate responsibilities as a significant part of the electric industry.

"We believe this new supplemental financing will go a long way in solving the problem of securing new capital for plant.

"Wholesale power supply is just as crucial to the objective of serving our members with an adequate supply of power in keeping with good business practices. It is our stated objective in the rural electric program to achieve effective influence, control, or ownership of an assured and adequate source of wholesale power in order to provide low cost total utility service in our operating areas.

"Our third consideration is to maintain territory integrity. As rural electric systems we have worked to develop what was once unserved rural areas of the nation. The profile of rural America has changed and electric power supplied by rural electric systems has contributed to that change. This is where our member-owners exist—it shall be our purpose to continue serving these same areas that we have served for the past thirty years, although they are not in some instances altogether rural today.

"Rural electric systems must achieve and maintain territorial protection to assure the continued development of economically sound systems, able to adequately serve all present and future electric power requirements in our service area.

"We are living in a period of rapid change. We must accept it and whenever possible, lead it. If we are sincere in our efforts to move with the tide of change, we may be required to change some of our operating philosophies and to lay aside some of our cherished traditions and concepts. Perhaps former Federal Power Commission Chairman Lee White said it best when he stated, 'Today the electric utility industry—

and that term includes every segment of the industry—is truly facing a future unrecognizable in terms of the past and even the present.'

"None of us know what the future may bring, but we prefer to think of our present as a time of greatest challenge. We are willing to cooperate with all segments of the industry and with any others who are charged with the responsibility of meeting these challenges. Never before has it been more necessary for all of us in the electric utility business to work together to meet common problems.

"As former TVA Chairman David Lilienthal put it, 'Unless the business of supplying people with reasonably priced, reliable electric power is carried on by private corporations with a due sense of responsibility to the paramount public interest...the public at any time may assure the function of providing itself with this necessity of community life.'

"In reality, this is a statement of our basic challenge. We must meet this challenge; otherwise, it will be an open invitation for others to step in and attempt to do the job for us. This is a development we must avoid at all cost.

"But if we work together in mutual understanding for the common good, then we need have no doubts or fears of the future. As Franklin D. Roosevelt once stated, 'The only limit to our realization of tomorrow will be our doubts of today. Let us move forward with strong and active faith.' If we will move forward together, we can make this time of our greatest challenge, our time of greatest opportunity."

ACKNOWLEDGMENTS

Let There Be Light would not have been possible without the cooperation of the CEOs, Presidents, Managers, and consumer-owners of the present twenty-six Kentucky rural electric cooperatives. All those who took the time to share their memories of how it was before the lights went on spoke from their hearts in a spirit of cooperative accord.

J.K. Smith, the Kentucky Association of Electric Cooperatives visionary, was generous throughout.

Ron Sheets, President of KAEC, was supportive, start to finish. His suggestions were always timely, never intrusive.

Dennis Cannon, KAEC Vice President for Member and Public Relations, was tireless in the long search through archives, where words and pictures lay hidden. His cooperative counterparts across the Commonwealth were unfailingly helpful.

Paul Wesslund, KAEC Vice President and Editor of *Kentucky Living* magazine, and Anita Travis, Managing Editor of *Kentucky Living*, were patient stewards of my monthly column, "The View from Plum Lick." Former editor Gary Luhr conceived the idea of the "back page," planting the seed that would grow to become *Let There Be Light*.

Barbara Rodgers, Director of KAEC Program Coordination, was an invaluable help in bringing together the research bits and pieces.

I commend Paula Sparrow, Quality Control Specialist for *Kentucky Living*, for her exceptional "co-op family" proofreading detail.

As always, my thanks to editor Georgiana Strickland for painstaking attention to accuracy and usage. Graphic Designer Stacey Freibert, too, worked diligently under deadline pressures to complete the volume.

But without my soul mate, Eulalie ("Lalie"), there would be no book. Her editing and encouragement were irreplaceable; it was she who drove me to each of Kentucky's electric co-ops, while making sure my doctor's appointments were unfailingly kept at M.D. Anderson Cancer Center in Houston, Texas.

Dr. Paul Mathew and his associates kept me alive, holding prostate cancer at bay, and no amount of my gratitude will be enough.

Thanks to Percival Beacroft and Troy Beacroft for making their Texas cabin available to us during the early treatment days. Our gratitude also goes to Lalie's sisters, Cornelia Conques and Charlotte Holloman, for opening the doors of the homeplace in Mississippi, where convalescence continued among family and friends.

Through it all, *Let There Be Light* has been among the best of medicines, and with it I acknowledge responsibility for each word of a story aching to be told.

David Dick
Plum Lick, Kentucky
November 1, 2008